INDIVIDUALISED MATHEMATICS

Developed by the School Mathematics Project in
association with the National Extension College

ALGEBRA 2
EQUATIONS, FORMULAS A...

CAMBRIDGE UNIVERSITY PRESS

Cambridge
London New York New Rochelle
Melbourne Sydney

The School Mathematics Project

When the SMP was found in 1961, its main objective was to devise radically new secondary school mathematics courses to reflect, more adequately than did the traditional syllabuses, the up-to-date nature and usages of mathematics.

SMP Books 1–5 form a five-year course leading to the O-level examination in SMP Mathematics. *Revised Advanced Mathematics Books 1, 2 and 3* cover the syllabus for the A-level examination in SMP Mathematics. Five shorter texts cover the material of the various sections of the A-level examination SMP Further Mathematics. There are two books for SMP Additional Mathematics at O-level. All the SMP GCE examinations are available to schools through any of the GCE examining boards.

Books A–H cover broadly the same development of mathematics as the first few books of the O-level series. Most CSE boards offer appropriate examinations. In practice, this series is being used very widely across all streams of comprehensive schools and its first seven books, together with *Books X, Y and Z*, provide a course leading to the SMP O-level examination. *SMP Cards I and II* provide an alternative treatment in card form of the mathematics in *Books A–D*. The six units of *SMP 7–13*, designed for children in that age-range, provide a course for middle schools which is also widely used in primary schools and the first two years of secondary schools. Teacher's Guides accompany all these series.

The SMP has produced many other texts, and teachers are encouraged to obtain each year from the Cambridge University Press, P.O. Box 110, Cambridge CB2 3RL, the full list of SMP publications currently available. In the same way, help and advice may always be sought by teachers from the Executive Director at the SMP Office, Westfield College, Kidderpore Avenue, London NW3 7ST. The Annual Reports, details of forthcoming in-service training courses and other information may be obtained from the SMP Office.

The SMP is continually evaluating old work and preparing for new. The effectiveness of the SMP's work depends, as it has always done, on the comments and reactions received from teachers and pupils in a wide variety of schools using SMP materials. Readers of the texts can, therefore, send their comments to the SMP in the knowledge that they will be taken into consideration.

The authors of the original books on whose work this series is based are named in *The School Mathematics Project: The First Ten Years*, published by the Cambridge University Press.

SMP Individualised Mathematics has been produced by a team consisting of

Judy Bonsall   G. Merlane
G. S. Howlett   L. Savins
M. K. Leach   D. R. Skinner
J. L. Lloyd   J. V. Tyson

John Lloyd led the work on the series until his death in 1977, and the final editing has been carried out by Derek Skinner. Many others have helped with advice and criticism, particularly those students who worked through the material in draft form.

Contents

			page
		Preface	v
		How to use this book	vi
1		**Equations and Orderings**	1
		Objectives	1
		Pre-test	1
	1.1	Simple equations: solution by inverse operations	1
	1.2	The use of inverse elements	5
	1.3	One-dimensional orderings	8
	1.4	Equations in two unknowns	11
	1.5	Orderings in two unknowns	15
		Summary	18
		Post-test	19
		Assignment	20
		Answers	20
2		**Formulas**	29
		Objectives	29
		Pre-test	29
	2.1	Construction of formulas	29
	2.2	Calculations using formulas	31
	2.3	Changing the subject of a formula	32
	2.4	Further examples	34
		Summary	35
		Post-test	36
		Assignment	36
		Answers	37
3		**Gradients**	41
		Objectives	41
		Pre-test	41
	3.1	Rates of change	42
	3.2	Gradients	44
	3.3	Gradient of a linear function	48
	3.4	Gradient of a non-linear function	49
	3.5	Distance, speed and acceleration	51
		Summary	54
		Post-test	55
		Assignment	57
		Answers	57

4	**Linear programming**	73
	Objectives	73
	Pre-test	73
4.1	Identifying the problem	74
4.2	Graphical representation	76
4.3	Maximising and minimising	78
	Summary	79
	Post-test	80
	Assignment	81
	Answers	81

Preface

SMP Individualised Mathematics is based upon the content of *SMP Books 1–5* and *Books A–G, X, Y, Z*, covering the syllabus for the O-level SMP Mathematics.

There are two main distinguishing features of the series. First, the material is presented in a programmed form and the books are thus suitable for use in individualised learning, where self-assessment and clear explanation play a major role. The carefully structured development of each topic makes the books suitable for use by students working alone with minimum tuition, in schools, technical colleges, colleges of further education and on courses organised by the National Extension College.

Secondly, instead of the spiral development of the SMP texts, *SMP Individualised Mathematics* presents the mathematics by topics. Each book, apart from the two devoted to revision, presents the work on a particular theme. Hence the books will prove useful to pupils who have missed work through absence from class, to students coming from abroad, and to pupils transferring to a different school. The style and arrangement of these books should make them very suitable for use by pupils in the sixth form who are working to improve their earlier performance at CSE or O-level. The books will also be useful for revision and consolidation.

Although written with the SMP O-level course in mind, *SMP Individualised Mathematics*, like other SMP texts, can be used to prepare for other O-level examinations based on similar syllabuses.

The titles in this series are as follows:
1. Computation and Graphs
2. Probability and Statistics
3. Algebra 1: Language and Structure
4. Algebra 2: Equations, Formulas and Graphs
5. Further Algebra and Computation
6. Matrix Algebra and Isometric Transformations
7. Further Matrices and Transformations
8. Geometry 1: Symmetry and Trigonometry
9. Geometry 2: Shapes and Similarity
10. Geometry 3: Three Dimensions
11. Revision 1
12. Revision 2

How to use this book

Each chapter begins with a list of what you should be able to do after studying the chapter. This is followed by a pre-test, which gives you some idea of what you should know before you start that particular chapter. If you have difficulty with the pre-test, you should revise the work required for it – from either the appropriate chapter of this or a companion book, or an elementary text-book – before continuing with the chapter.

The teaching part of the chapter is divided into several sections, and includes a number of exercises. Other questions are asked in the text, and *you should write down the answers to all these questions and exercises in a notebook* as you go along. The start of each set of questions is marked by a white triangle on the left-hand side of the page. When you come to a triangle with a number in it (on the *right-hand* side of the page) you should check your work to that point by turning to the answers at the end of the chapter and finding the triangle with the same number (now on the *left-hand* side of the page).

The teaching part of the chapter is followed by a summary of the important results of the chapter (you may well find it helpful to copy these into a separate notebook that is kept especially for revision), and a post-test to test your understanding of the chapter as a whole. The answers to this post-test are also at the end of the chapter.

Finally (apart from the answers) there is an 'assignment'. This is another exercise covering the whole chapter, but this time there are no answers in this book. If at all possible you should have *this* exercise marked by a teacher or tutor.

1 Equations and orderings

Objectives

This is what you should be able to do after studying this chapter.
(1) Solve a simple equation by using inverse operations and inverse elements.
(2) Solve one-dimensional orderings, and simultaneous one-dimensional orderings, and show the solutions on a number line.
(3) Represent graphically equations and orderings in two unknowns.
(4) Solve simultaneously two equations in two unknowns, both algebraically (using either substitution or elimination) and graphically.
(5) Show graphically the solution set of two or more orderings in two unknowns, considered simultaneously.

Pre-test

1. Find the answers to the following sums.
   (a) $2 + {}^-4$ (b) $2 - {}^-5$ (c) $\frac{3}{4} + \frac{1}{5}$ (d) $2\frac{1}{2} - 1\frac{2}{3}$ (e) ${}^-2 \times {}^-\frac{1}{2}$ (f) $\frac{3}{4} \div {}^-\frac{1}{6}$

2. Find x in the following cases.
   (a) $2x = {}^-7$ (b) $x - 4 = {}^-\frac{1}{2}$ (c) $\frac{2}{3}x = \frac{5}{3}$ (d) ${}^-\frac{1}{7}x = {}^-7$

3. Rewrite the following without using brackets.
   (a) $2(x+3)$ (b) $\frac{1}{2}(x-6)$ (c) ${}^-4(6-x)$

4. Illustrate the following on number lines.
   (a) $x = 3$ (b) $x \leqslant 3$

5. Draw graphs of the following.
   (a) $2x + y = 2$ (b) $3x - 4y = {}^-12$

1.1 Simple equations: solution by inverse operations

Using flow diagrams

Take a number (call it x), multiply it by 3, and add 2, and the answer is 14. We can represent the information given in this statement by the flow diagram

$x \longrightarrow \boxed{M\ 3} \longrightarrow \boxed{A\ 2} \longrightarrow 14$

where the boxes represent the operations that we are asked to perform. [*Note:* In this chapter we shall use A for 'add', S for 'subtract', M for 'multiply by' and D for 'divide by'.]

Write down the flow diagrams for the following.

▷ 1   Take a number, subtract 7, divide by 5, and the answer is 1.

2   Take the number 4, add 6, divide by 4, subtract 4 and the answer is x.

Does it make any difference if we perform the operations in a different order? Write down the flow diagrams for the following.

3   Take a number, divide by 5, subtract 7, and the answer is 1.

4   Take the number 4, subtract 4, divide by 4, add 6 and the answer is x.

Write down the values of x in **2** and **4**, and guess (or find by trial and error) values of x for the above example and **1** and **3**. Does the order of the operations matter? ▷ 2

The flow diagram for the above example can be written

$$x \xrightarrow{x} \boxed{M\ 3} \xrightarrow{3x} \boxed{A\ 2} \xrightarrow{3x+2} 14$$

which gives the relation $3x+2 = 14$.

In the same way **1** can be written as

$$x \xrightarrow{x} \boxed{S\ 7} \xrightarrow{x-7} \boxed{D\ 5} \xrightarrow{(x-7)/5} 1$$

giving the relation $\dfrac{x-7}{5} = 1$.

▷ 5   Obtain the relation for **3**. ▷ 3

These relations contain an 'equals' sign, they contain at least one letter (usually x) representing some number whose value is not known to begin with, and they are correct for only a few values of x. Such relations are called *equations*. If only one letter (say x) is involved they are 'equations in one unknown', if they are correct for only one value of x they are called 'simple equations'. The value(s) of x which give correct statements are known as the 'solutions' (or 'roots') of the equation, and the process of finding these values is known as 'solving the equation'. (That is, an equation is considered as solved when it is rearranged in the form $x = \ldots$.)

Exercise A

▷ 1   Write down the result of carrying out the instructions given in the following flow diagrams.

(a) $3 \longrightarrow \boxed{A\ 7} \longrightarrow \boxed{M\ 4} \longrightarrow \boxed{S\ 3} \longrightarrow$

(b) $\tfrac{3}{4} \longrightarrow \boxed{D\ ^-3} \longrightarrow \boxed{S\ \tfrac{1}{2}} \longrightarrow$

(c) $6 \longrightarrow \boxed{S\ 2} \longrightarrow \boxed{M\ ^-\tfrac{1}{2}} \longrightarrow$

(d) $4 \longrightarrow \boxed{D\ ^-2} \longrightarrow \boxed{S\ ^-3} \longrightarrow \boxed{M\ ^-1} \longrightarrow$

2 Write down the equations represented by the following flow diagrams.
  (a) $x \longrightarrow \boxed{A\ 2} \longrightarrow \boxed{M\ 3} \longrightarrow 14$
  (b) $x \longrightarrow \boxed{S\ 2} \longrightarrow \boxed{D\ 2} \longrightarrow \boxed{A\ 2} \longrightarrow 3$
  (c) $x \longrightarrow \boxed{M\ ^-2} \longrightarrow \boxed{A\ 4} \longrightarrow \boxed{M\ \tfrac{1}{2}} \longrightarrow 3$
  (d) $x \longrightarrow \boxed{D\ 3} \longrightarrow \boxed{A\ 4} \longrightarrow \boxed{D\ 2} \longrightarrow 1$

3 Construct the flow diagrams corresponding to the following equations.
  (a) $4(x+3) = 6$
  (b) $2x+3 = 6$
  (c) $(x-4)/3 = 5$
  (d) $\dfrac{5x+1}{3} = 12$

Reversing the flow diagram

1 Complete the following four flow diagrams.
  $x \longrightarrow \boxed{A\ 6} \longrightarrow \boxed{S\ 6} \longrightarrow$
  $x \longrightarrow \boxed{S\ 9} \longrightarrow \boxed{A\ 9} \longrightarrow$
  $x \longrightarrow \boxed{M\ 5} \longrightarrow \boxed{D\ 5} \longrightarrow$
  $x \longrightarrow \boxed{D\ 5} \longrightarrow \boxed{M\ 5} \longrightarrow$

You should notice that in each case the two operations together leave x unchanged. In each case we say that the second operation is the inverse of the first operation. (You should be able to see that the actual numbers used in the four examples above could have been replaced by any others, so long as the same number is used for each pair of operations.)

We can now make use of this idea to 'run' our flow diagram backwards; for example, for the flow diagram

$x \longrightarrow \boxed{M\ 3} \longrightarrow \boxed{A\ 2} \longrightarrow 14$
$4 \longleftarrow \boxed{D\ 3} \overset{12}{\longleftarrow} \boxed{S\ 2} \longleftarrow 14$

'undoes' the statement, or works the reversed flow diagram backwards, step by step. (Remember that if you put *on* your socks and then put *on* your shoes, the reverse process is to take *off* your *shoes* first, and then take *off* your *socks*.) The reversed flow diagram shows us that we must have started with the number 4, and so the solution to the equation $3x+2 = 14$, set up by the flow diagram in the first line, is 4, or $x = 4$.

Similarly the reversed flow diagram for

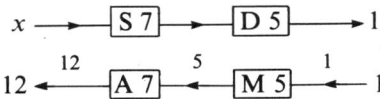

is

giving $x = 12$ as the solution to the equation $(x-7)/5 = 1$.

2 Reverse the flow diagram for **3** in the section 'Using flow diagrams.' For what equation is this the solution?

3

Exercise B

 1 Find, using the reversed flow diagram method, the values of x that satisfy the equations in questions **2** and **3** of Exercise A.

2 Solve the equations
(a) $6(x-5) = 3$ (b) $\frac{1}{2}x + 4 = 2$ (c) $3(4+x) = 6$ (d) $\frac{x+7}{3} - 1 = 2$

Self-inverse operations

The operations or instructions $\boxed{M\ 3}$, $\boxed{A\ 2}$ etc. are, of course, simple examples of functions. They could have been written

$$x \to 3x, \qquad x \to x+2 \qquad \text{etc.}$$

A chain of instructions such as $\to \boxed{M\ 3} \to \boxed{A\ 2} \to$ is the composition of two functions, and since $\boxed{D\ 3}$ and $\boxed{S\ 2}$ are the corresponding inverse functions $\to \boxed{S\ 2} \to \boxed{D\ 3} \to$ is the inverse of the composite function.

The other two types of function that we need to consider at this stage are the self-inverse functions of the forms

$$x \to 4-x \quad \text{and} \quad x \to \frac{12}{x}.$$

(We could, of course, replace the numbers 4 and 12 by any other numbers without altering the self-inverse character of these functions.)

 1 If $f: x \to 4-x$ and $g: x \to 12/x$ wrote down the values of the following.
(a) $f(3)$ (b) $ff(3)$ (c) $g(3)$ (d) $gg(3)$

2 You should also be able to write down the values of the following without doing any working.

(a) $ff(5)$ (b) $ff(\bar{\ }2)$ (c) $gg(5)$ (d) $gg(\bar{\ }2)$

In this chapter we shall label such functions as $\boxed{\text{S from 4}}$ for 'subtract from 4' and $\boxed{\text{D into 12}}$ for 'divide into 12'.

(The function $\boxed{\text{D into 1}}$ is the reciprocal function $x \to \frac{1}{x}$, and as such is tabulated in many books of tables. In numerical work it is often convenient to consider $\boxed{\text{D into 12}}$ as itself a composition of the two functions $\boxed{\text{D into 1}}$ and $\boxed{\text{M 12}}$.)

Now consider the solution of the equation

$$2\left(4 - \frac{6}{x}\right) = 12$$

The flow diagram for this is

$$x \to \boxed{\text{D into 6}} \to \boxed{\text{S from 4}} \to \boxed{\text{M 2}} \to 12$$

and so the reversed flow diagram is

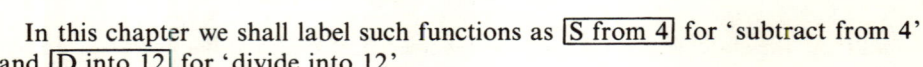

4

which can be written from left to right as

$$12 \xrightarrow{12} \boxed{D\ 2} \xrightarrow{6} \boxed{S\ \text{from}\ 4} \xrightarrow{-2} \boxed{D\ \text{into}\ 6} \xrightarrow{-3} -3$$

giving the solution $x = {}^{-}3$.

Exercise C

Solve the following equations.

1. $2x - 3 = 6$
2. $3 - 2x = 6$
3. $15 - 2x = 8$
4. $12 - \frac{1}{2}x = 7$
5. $9 - x = {}^{-}5$
6. $x/6 + 3 = 6$
7. $6/x + 3 = 6$
8. $6/(x+3) = 6$
9. $\frac{1}{2}(x-2) = 7$
10. $\frac{1}{2}(2-x) = 7$
11. $12/(4-x) = 3$
12. $4 - 12/x = 3$

1.2 The use of inverse elements

Inverse elements

The use of flow diagrams and inverse operations is a good method for solving a wide variety of simple equations in one unknown. However, there are some restrictions, and so we are now going to consider another method for solving equations such as

$$2x + 3 = 4 - x \quad \text{and} \quad 2(x-4) = 3(2-5x)$$

Two ideas are involved in this method:

(1) We can consider an equation as two sides that balance each other about the equals sign as shown in Figure 1.

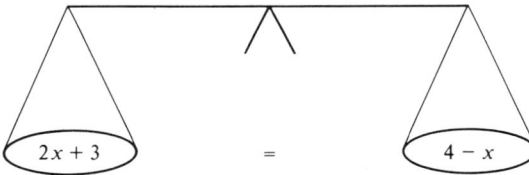

Figure 1

We can do anything we like (within reason) to one side of this equation, and still maintain the balance, *provided we do exactly the same thing to the other side.*

Thus, for example, we can add 6 to both sides, subtract 4 from both sides, subtract both sides from 10, multiply both sides by 7, multiply both sides by ${}^{-}1$, divide both sides by 5, divide both sides into 12, take the positive square root of both sides, take the cube of both sides, take the reciprocal of both sides etc.

(2) We can use our knowledge and ideas of identity and inverse elements for situations of the form $a*b = c$, where $*$ represents *any* binary operation, and apply them specifically to the solutions of the equations we are considering in this chapter.

In particular, if ∗ is addition, then the identity element is 0 (sometimes called the Zero Element) because for any number n

$$n+0 = 0+n = n$$

and so the inverse of n under addition is ^-n, because

$$n+{^-n} = 0 \quad \text{(the Identity Element)}.$$

If ∗ is multiplication, then the identity is 1 (sometimes called the Unit Element), and the inverse of n under multiplication is $1/n$. (The numbers ^-n and $1/n$ are sometimes called respectively, the additive and the multiplicative inverses of n.)

In applying ideas (1) and (2) to solve equations by using inverse elements we restrict ourselves to the operations of addition and multiplication. In order for this to be possible it is sometimes necessary to regard such statements as $2x-3$ as $2x+{^-3}$. The method is best illustrated by some examples.

Example 1 (treating the equation at the beginning of this chapter by this method)

	To solve	$3x+2 = 14$
Add the additive inverse of 2 to both sides:		$(3x+2)+{^-2} = 14+{^-2}$
Apply the associativity law for addition:		$3x+(2+{^-2}) = 12$
i.e. since *element* ∗ *inverse* = *identity*:		$3x+0 = 12$
i.e. since *element* ∗ *identity* = *element*:		$3x = 12$
Multiply both sides by the multiplicative inverse of 3:		$\tfrac{1}{3} \times 3x = \tfrac{1}{3} \times 12$
Apply the associativity law for multiplication:		$(\tfrac{1}{3} \times 3) \times x = 4$
i.e. since *element* ∗ *inverse* = *identity*:		$1 \times x = 4$
i.e. since *element* ∗ *identity* = *element*:		$x = 4$

which is the solution.

Example 2   To solve $4x-5 = 7$
Rewrite as: $4x+{^-5} = 7$
Add $^+5$ to both sides (as $^-5+{^+5} = 0$, the identity): $4x+{^-5}+{^+5} = 7+{^+5}$
which simplifies to: $4x = 12$
giving a solution of: $x = 3$

Example 3

$$\frac{x+6}{8} = {^-3}$$

Rewrite as $\tfrac{1}{8}(x+6) = {^-3}$
multiply by 8 $(x+6) = {^-24}$
add $^-6$ $x = {^-30}$

Exercise D

Use the method of inverse elements to solve the following equations.

▷ 1 $7x+5 = 8$ 4 $3(2x-1) = 8$
2   $5x-6 = 14$ 5 $\tfrac{3}{8}x+\tfrac{1}{4} = \tfrac{1}{8}$
3   $2-3x = 26$ 6 $\dfrac{x+2}{3} = 4$

Collecting terms

At the beginning of the previous section on 'Inverse elements', two equations were given. The first was
$$2x+3 = 4-x$$
This is rewritten as
$$2x+3 = 4+{}^-x$$
We now have an equation in which the unknown x appears on both sides. In cases like this we first 'collect' these terms on the same side. We usually like to have the x-term(s) on the left-hand side of an equation, and so in this example we shall 'remove' the x-term from the right-hand side by adding its inverse (^+x) to both sides, giving
$$2x+3+{}^+x = 4+{}^-x+{}^+x$$
which simplifies to $\quad 3x+3 = 4$

and then $\quad 3x = 1$ (adding $^-3$ to both sides)

giving $\quad x = \tfrac{1}{3}$ (multiplying both sides by $\tfrac{1}{3}$).

Checking the solution

A common method of checking the answer to an equation is to replace x (by the value we have found for the solution) in both sides of the *original* equation independently.
To check that $x = \tfrac{1}{3}$ is a solution to the equation $2x+3 = 4-x$.
When $x = \tfrac{1}{3}$, the left-hand side (LHS) has a value of $2(\tfrac{1}{3})+3 = 3\tfrac{2}{3}$.
Similarly, when $x = \tfrac{1}{3}$, the RHS $= 4-(\tfrac{1}{3}) = 3\tfrac{2}{3}$.
Hence, when $x = \tfrac{1}{3}$, the LHS = the RHS ($= 3\tfrac{2}{3}$), and so $x = \tfrac{1}{3}$ *is* a solution of the equation.

Using the distributive law

The other equation at the beginning of the section 'Inverse elements' was
$$2(x-4) = 3(2-5x)$$
The terms containing x are now inside the brackets. The equation can be rewritten in a workable form by using the 'distributive law for multiplication over addition', i.e. $a \times (b+c) = a \times b + a \times c$. Thus we have
$$2x-8 = 6-15x \text{ ('removing brackets')}$$
or $\quad 2x+{}^-8 = 6+{}^-15x$

giving $\quad 2x+{}^-8+{}^+15x = 6+{}^-15x+{}^+15x$ (adding ^+15x to both sides)

or $\quad 17x+{}^-8 = 6$

which gives $\quad 17x = 14$ (adding $^+8$ to both sides)

and so $\quad x = \tfrac{14}{17}$ (multiplying both sides by $\tfrac{1}{17}$)

(Note that when we write $2x+15x = 17x$ we are also using the distributive law in the form $a \times b + a \times c = a \times (b+c)$, together with the commutative law.)

Graphical solutions

Equations in one unknown can also be solved by graphical methods. These are particularly useful for more complicated equations which may have two or more solutions. The method will be mentioned briefly later in this chapter, and is developed more fully in the chapter on quadratic functions in *Further Algebra and Computation*.

Exercise E

1. Use the distributive law to simplify the following.
   (a) $\frac{1}{6}x + \frac{1}{7}x$ and (b) $\frac{1}{4}x + \frac{1}{4}y$
   and to expand (c) $3(5-2x)$ and (d) $^-4(3-2x)$

2. Solve, by the use of inverse elements etc., and check, the following equations.
   (a) $5x+6 = 3x+8$
   (b) $\dfrac{5-2x}{4} = 7x$
   (c) $\dfrac{x+1}{4} = \dfrac{x-2}{3}$
   (d) $1 - \dfrac{5}{x} = 6$
   (e) $\dfrac{5}{x+3} = 4$
   (f) $3 - \dfrac{x+1}{2} = \dfrac{x-1}{4} + 1$
   (g) $5(2x-1) = 2+4x$
   (h) $\frac{1}{2}(1-3x) = \frac{3}{4}(2x+20\frac{2}{3})$

1.3 One-dimensional orderings

The statement or relation $2x+3 > 9$ is not an equation as it does not contain an equals sign. It is called an *inequation*, *inequality* or *ordering*. (The last name is used because the relations $>$ and $<$ arrange numbers in order of size.) Just as an equation can be represented as a balance between the two sides, an inequality can be represented by an 'imbalance' between the two sides as shown in Figure 2.

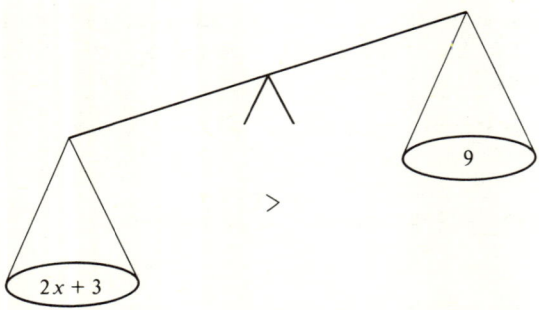

Figure 2. $(2x+3)$ is heavier than (9)

It should therefore be clear that we do not change the imbalance if we add the same thing to both sides, subtract the same thing (or, for example, add ⁻5) from both sides, or multiply both sides by the same positive number.

For example, what happens to the inequality $3 < 5$ if we do the following?

1. Add 7 to both sides.
2. Add ⁻6 to both sides.
3. Subtract 1 from both sides.
4. Multiply both sides by 4.
5. Multiply both sides by $\frac{3}{4}$.

(Remember that $x < y$ means that x is to the left of y on the number line

```
 ++++++++++++++++++++++++++++++
 ⁻10     ⁻5    x   0      5    y   10
```

or that x is below y on a 'thermometer'. Thus $⁻8 < ⁻5$ etc.)

But what happens to the imbalance if we multiply both sides by ⁻2? By performing this on the inequality $3 < 5$, we see that if we multiply both sides by a *negative* number we change the direction of an ordering. It is important to remember this! Bearing this last point in mind, the solution of an inequality by using inverse elements is very similar to solving an equation. Some worked examples will show this.

Example 1
$$2x+3 > 9$$
$$2x > 6 \text{ (adding } ⁻3 \text{ to both sides)}$$
$$x > 3 \text{ (multiplying both sides by } \tfrac{1}{2}\text{).}$$

Example 2
$$5-4x < 12$$
Rewrite as
$$5 + ⁻4x < 12$$
$$⁻4x < 7 \text{ (adding } ⁻5 \text{ to both sides)}$$
$$x > ⁻\tfrac{7}{4} \text{ (multiplying both sides by } ⁻\tfrac{1}{4}\text{).}$$

(Note in this last stage that the multiplicative inverse of ⁻4 is ⁻$\tfrac{1}{4}$, and, as we are multiplying both sides of the inequality by a negative number, the 'direction' of the inequality has to be reversed.)

Example 3
$$3(x+4) \geqslant 5x-8$$
or
$$3(x+4) \geqslant 5x + ⁻8$$
$$3x+12 \geqslant 5x + ⁻8 \text{ (removing brackets)}$$
$$⁻2x+12 \geqslant ⁻8 \text{ (adding } ⁻5x \text{ to both sides)}$$
$$⁻2x \geqslant ⁻20 \text{ (adding } ⁻12 \text{ to both sides)}$$
$$x \leqslant 10 \text{ (multiplying both sides by } ⁻\tfrac{1}{2}\text{).}$$

Exercise F

Solve the following using the method of inverse elements.

1   $x+4 > {}^-2$ 5 ${}^-2x+3 \geqslant 7$

2   ${}^-2x > 10$ 6 $2(x+3) < 10$

3   $1-4x < 6$ 7 $\dfrac{x-4}{5} > {}^-1\tfrac{2}{5}$

4   $1 \leqslant 4-x$ 8 $4(x-1) \leqslant 3(x+5)-12$

Simultaneous one-dimensional orderings

What is implied by the statement $4 < x < 6$? The first part, $4 < x$, states that '4 is less than x', i.e. x is a number greater than 4. The second part, $x < 6$, states that x is a number less than 6. The combined statement, therefore, is a way of saying that x is a number between 4 and 6, or $\{x\}$ is the set of numbers that are greater than 4 and less than 6.

This last form suggests that $\{x\}$ may be considered as the intersection of the sets $\{x : x > 4\}$ and $\{x : x < 6\}$, i.e. the two inequalities $x > 4$ and $x < 6$ have to be satisfied *simultaneously* (i.e. at the same time).

We can illustrate this on the number line

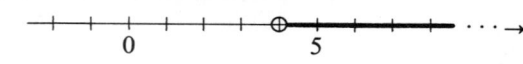

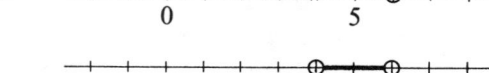

$4 < x < 6$

$\cdots\rightarrow$ implies that the set continues indefinitely in the direction of the arrow. The open circles at the ends of the lines show that x cannot be equal to 4 or to 6, but can take values as close to them as we like.

The inequality $4 < x \leqslant 6$ would be shown as

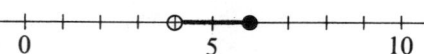

the closed circle showing that x *can* take the value 6. (Note that we usually write $x > 4$ rather than $4 < x$ following the convention of having x on the left-hand side of a statement, but we write $4 < x < 6$ in preference to $6 > x > 4$ to agree with the convention that the number line goes from left to right.)

How would we 'solve' a pair of simultaneous inequalities such as

$$6x - 1 > 1 - 2x > {}^-3?$$

By comparison with the previous example, the solution of this is the intersection of the sets $\{x : 6x-1 > 1-2x\}$ and $\{x : 1-2x > {}^-3\}$.

The first of these is $6x - 1 > 1 - 2x$

or $6x + {}^-1 > 1 + {}^-2x$

10

add ⁺2x to both sides	$8x + {}^-1 > 1$
add ⁺1 to both sides	$8x > 2$
multiply both sides by $\frac{1}{8}$	$x > \frac{1}{4}$

Similarly the second inequality simplifies to $x < 2$ and so the solution to the simultaneous inequalities is $\frac{1}{4} < x < 2$.

Exercise G

1. Illustrate the following on a number line.
   (a) $^-2 < x < 3$
   (b) $^-4 < x \leqslant 0$
2. Does $^-4 > x \geqslant 0$ have any solution?
3. Solve, where possible, the following simultaneous inequalities.
   (a) $3x > x+2 > 2x-4$
   (b) $4+x < 1 < 2x$
   (c) $4+x > 1 \geqslant 2x$
   (d) $5-x < 4 < 1+x$

1.4 Equations in two unknowns

Statements such as $x = 3$ and $x+3 > 7$, which involve just one unknown (or variable) x, can be represented on a number line, and are therefore said to be 'one dimensional'.

Statements such as $x+2y = 7$ and $y \geqslant 2x$ involve two variables (usually x and y) and therefore need two dimensions to represent them, such as the familiar x, y graph.

Consider the 'equation' $x+2y = 7$. How many solutions does this equation have? What exactly do we mean by a 'solution' in this case? By inspection (i.e. intelligent trial and error!) we see that the equation balances for $x = 1$ and $y = 3$, or for $x = 3$ and $y = 2$, or for $x = 5$ and $y = 1$ etc. Thus the 'ordered pairs', (1, 3), (3, 2), (5, 1) etc., are all solutions of the equation. In fact, we can choose *any* value for y (or x)

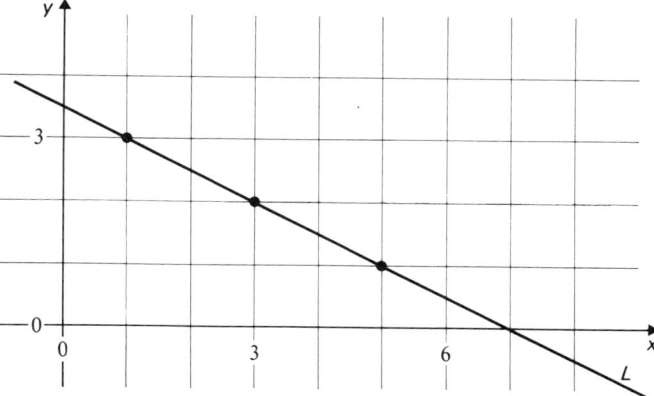

Figure 3

and find *one* associated value of x (or y) such that the equation balances. Hence, there is an infinite number of 'solutions' of this equation. If we plot all these solutions as coordinates on an x, y graph we obtain a straight line L as shown in Figure 3.

Every possible solution of the equation is represented by a point somewhere on this line, and every point of the line represents a possible solution of the equation: we say that '$x+2y = 7$ is the equation of the line L'.

Simultaneous equations in two unknowns

Each of the equations $x+2y = 7$ and $y = 3x-7$ can be represented by a line on an x, y graph. Some points for the second equation can be found by completing the following table.

x	$y = 3x-7$
1	$3(1)-7 = {}^-4$
2	$3(2)-7 = 6-7 = {}^-1$
3	$3(3)-7 = 9-7 = 2$

Remember that although two points are sufficient to draw a straight line, it is safer to plot at least three; if you can't draw one straight line through all three points then you have made a mistake somewhere!

All the possible solutions of $x+2y = 7$ lie on the line L, and all the solutions of $y = 3x-7$ lie on the line M; if we want to solve the equations *simultaneously* we need the coordinates of a point that is on both the lines. Often there is just one such point (P in Figure 4). You may have used this graphical method for solving simultaneous equations before.

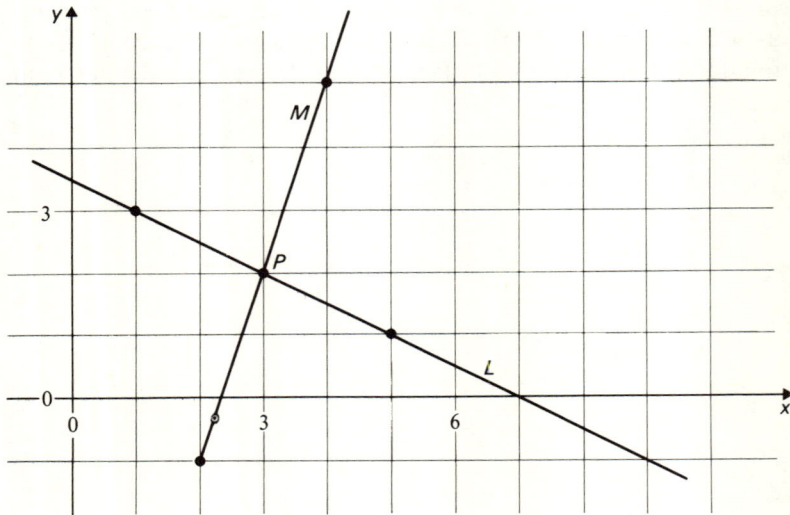

Figure 4

Exercise H

Solve the following pairs of simultaneous equations graphically.

1 $\begin{cases} x+y = 6 \\ y = 4x-4 \end{cases}$ 2 $\begin{cases} x = 3y-1 \\ 4y = 2x+3 \end{cases}$ 3 $\begin{cases} 2x+y = 1 \\ 2x+5y = 9 \end{cases}$

Algebraic methods

1 *Substitution*: (In dealing with simultaneous equations it is good practice to number the equations, as is done below.)

Again let us suppose that we wish to solve

and
(i) $x+2y = 7$
(ii) $y = 3x-7$

simultaneously.

Equation (ii) gives us an expression for y in terms of x. We can replace the y in equation (i) by this expression, to obtain (iii):

(iii) $x+2(3x-7) = 7.$

This can now be solved by the methods developed earlier in this chapter, to give the result $x = 3$.

This value of x is now substituted into one of the earlier equations; in this case it is easier to use equation (ii). Replacing x by the value 3 in this equation, we have

(iv) $y = 3(3)-7$

which simplifies to $y = 2$, and so the simultaneous solution of the two equations is $x = 3, y = 2$. This answer should be 'checked' (as always!). As we used equation (ii) to find the value of y, we check in equation (i) by replacing x and y by the values we have found.

Substituting 3 for x and 2 for y in (i)

the LHS $= 3+2(2) = 7$
the RHS $= 7$

and so the solution 'checks'.

This method is recommended only for those cases in which it is possible to find x in terms of y (or y in terms of x) easily and without involving too many fractions.

1 Solve the equations

(i) $x = 3y-1$
(ii) $4y = 2x+3$

(a) by substituting (from ii) $\frac{1}{4}(2x+3)$ for y in (i),
(b) by substituting $(3y-1)$ for x in (ii).
(c) Which is the easier solution? (Are they the same?)

2 *Elimination (or combination)*: Once again we shall consider the equations

(i) $x+2y = 7$
(ii) $y = 3x-7.$

For the elimination method it is best to express both equations with the x and y

terms on the left-hand side, with the x term first. Hence we shall need to rewrite (ii). Adding ^-3x to both sides we have (ii) as $^-3x+y = ^-7$.

Both (i) and (ii) may be pictured by balances, as in Figure 5, where in (i) $A = (x+2y)$ and $B = (7)$, and in (ii) $C = (^-3x+y)$ and $D = (^-7)$.

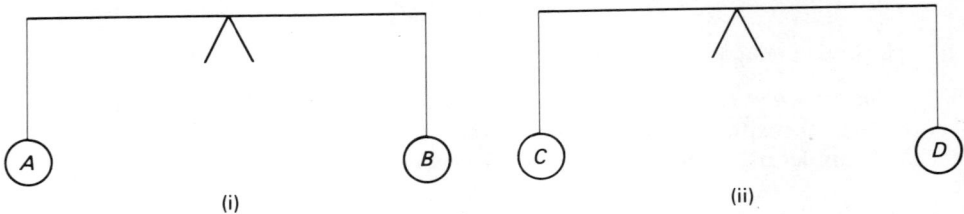

Figure 5

Since A balances B, and C balances D, any combination of A and C will balance the *same* combination of B and D. For example, $(A+3C)$ will balance $(B+3D)$, or $(2A-5C)$ will balance $(2B-5D)$ etc. The art of this method is to find a combination of A and C that gets rid of (i.e. eliminates) either x or y.

In this example the combination $(3A+C)$ eliminates x, and the combination $(A-2C)$ eliminates y. Using the first of these (a combination with a + is usually easier to use) we have

$$(3A+C) \text{ balances } (3B+D)$$

i.e.
$$3(x+2y)+(^-3x+y) = 3(7)+(^-7)$$

or
$$7y = 14$$

giving $y = 2$. To find the value of x we can now either use the other combination $(A-2C)$, or substitute 2 for y in one of the original equations. At this stage substitution is usually the easier method, and in this example substituting 2 for y in (i) gives $x+2(2) = 7$ or $x = 3$. (Having used (i) for finding x, we should now check by putting $x = 3$ and $y = 2$ in (ii).) In practice, this is set out as

(i)	$x+2y = 7$
(ii)	$^-3x+y = ^-7$
$3 \times$ (i)	$3x+6y = 21$
(ii)	$^-3x+y = ^-7$
3(i)+(ii)	$7y = 14$
multiply by $\frac{1}{7}$	$y = 2$
substitute in (i)	$x+4 = 7$
add $^-4$	$x = 3$.

Check, by substituting in (ii) as originally written
$$\text{LHS} = 2,$$
$$\text{RHS} = 3(3)-7 = 9-7 = 2.$$

As the LHS equals the RHS the solution has been checked.

3 *Matrices*: A third algebraic method is the use of matrices. This method is explained in the book *Further Matrices and Transformations*.

Graphical solution of equations in one unknown

The algebraic methods reduce a pair of simultaneous equations to one equation in one unknown, and we hope that we can solve that. But, since we have an alternative method of solving simultaneous equations (i.e. graphically), it seems possible that, if we have a 'difficult' equation in one unknown, we could 'retrace our steps', i.e. convert it into a pair of simultaneous equations, and solve them graphically. (In fact, we shall want only the solution for x.) For example, the equation $x+2 = \sqrt{(x-1)}$ could be the result of eliminating y from $y = x+2$ and $y = \sqrt{(x-1)}$.

Hence, if we draw the graphs of these two equations, the x-coordinate(s) of their point(s) of intersection will give the solution(s) of $x+2 = \sqrt{(x-1)}$.

This idea is developed in *Further Algebra and Computation*.

Exercise J

1. Solve the following by substitution.
   (a) $\begin{cases} x+2y = 10 \\ x-y = 4 \end{cases}$ (b) $\begin{cases} 2x+3y = 1 \\ 3x-y = 1 \end{cases}$

2. Solve the following by elimination.
   (a) $\begin{cases} x+y = 8 \\ x-y = 3 \end{cases}$ (b) $\begin{cases} 4x-3y = 1 \\ x-2y = 4 \end{cases}$ (c) $\begin{cases} 3x-2y = 4 \\ 2x+3y = {}^-6 \end{cases}$

1.5 Orderings in two unknowns

Two-dimensional orderings on a graph

The solution of the inequality

$$x+2y > 7 \quad \text{or} \quad \{(x, y): x+2y > 7\}$$

is the set of all points (x, y) for which the value of $(x+2y)$ is greater than 7. From the previous section we know that for all the points *on* the line L (see Figure 4) the value of $x+2y$ is equal to 7, so no point on that line is part of the solution. However, if we consider any point *above* the line (e.g. H in Figure 6), it has the

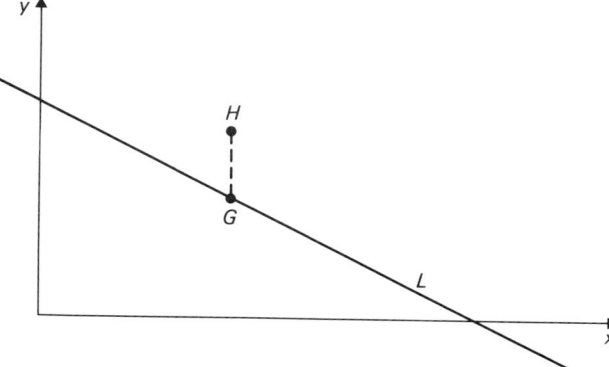

Figure 6

same x-coordinate as the corresponding point G on the line and a larger y-coordinate. For the point G the value of $x+2y$ is 7; hence for the point H the value of $x+2y$ is greater than 7, and so H (or more precisely the coordinates of H) is one solution of the inequality. But H could be *any* point above the line; therefore the solution of the inequality is the region above the line L. (Alternatively, we could have considered a point to the *right* of L.) Similarly, we could show that, for *all* points *below* L, $x+2y < 7$. Thus the region above the line is the complete solution of the inequality $x+2y > 7$.

If the solution of a single inequality is to be shown graphically it is quite common to show this by shading the required region (see Figure 7(a)). In this case the line L is drawn dotted, to show that it does *not* belong to the shading (points on L are not part of the solution set).

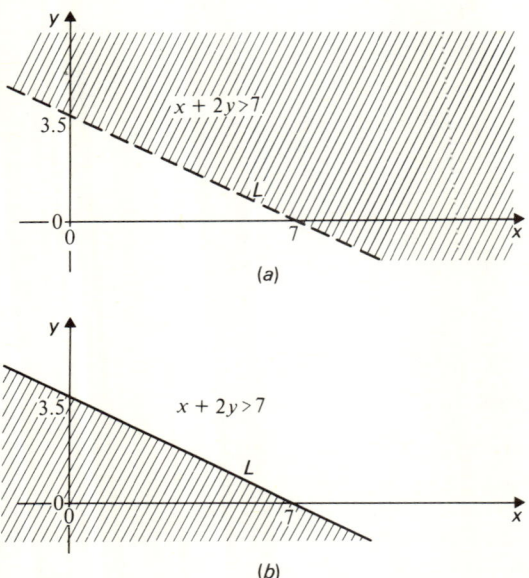

Figure 7

However, it is often more convenient, especially when showing two or more inequalities on the same graph, to shade *out* the *un*wanted region, as in Figure 7(b). The line L is now drawn as a continuous line to show that it *is* part of the shading. (As both conventions are in use, it is advisable to state on any work whether you have shaded in the required regions, or shaded out the unwanted regions.)

Simultaneous orderings

In this final section we consider the possible ways of solving a pair of inequalities simultaneously, first graphically, then algebraically. For our example we shall consider the orderings

(i) $x+y < 6$
(ii) $y < 4x-4$

1 *The graphical approach:* The boundaries of the two regions are the lines $x+y = 6$ and $y = 4x-4$. To decide which side of the boundary line is the required region, 'test' with one point. The origin (0, 0) is often very easy to use. Since $0+0 < 6$, the origin belongs to the region $x+y < 6$ and so *is* in (i). Hence we shade *out* the other side of the line $x+y = 6$.

Putting $x = 0$ and $y = 0$ in the two sides of (ii) we have LHS $= 0$, RHS $= 4(0)-4 = {}^-4$. Since $0 > {}^-4$, the origin lies in the region $y > 4x-4$, and so is *not* in (ii). Hence we shade out the side of the line $y = 4x-4$ that contains the origin.

To solve the orderings (i) and (ii) simultaneously, we want points that are in both regions, i.e. in the intersection of the two regions, and this is shown graphically in Figure 8 by the region that still has no shading in it. Neither boundary line is part of the solution set in this example.

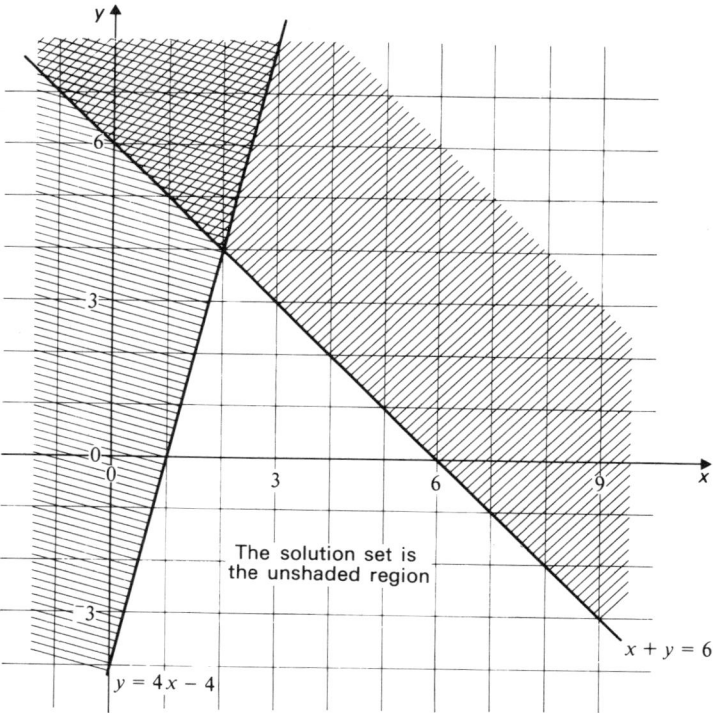

Figure 8. The unwanted regions are shaded out

2 *The algebraic approach:* At an elementary level there is not much that can be done algebraically with two or more inequalities, as trying to add or subtract inequalities is tricky!

> **1** (a) Try adding the following inequalities.

$$\begin{array}{cccc} 3>0 & 3>0 & 6>2 & 6>2 \\ \underline{2>1} & \underline{5>1} & \underline{3<5} & \underline{3<9} \end{array}$$

(b) Now try subtracting them. What do you notice?

17

In general we can say that if
$$A > B \quad \text{(think of this as 'A is heavier than B')}$$
and
$$C > D \quad \text{('C is heavier than D'),}$$
then
$$A+C > B+D, \quad A+3C > B+3D \quad \text{etc.}$$
and so
$$A-D > B-C, \quad A-3D > B-3C \quad \text{etc.}$$
but no more.

Now let us see what we can do with the two inequalities at the beginning of this section, where $A = (x+y)$, $B = (6)$, and C and D can be put in the forms $C = (^-4x+y)$ and $D = (^-4)$. The combination $4A+C$ gives us

$$4(x+y)+(^-4x+y) < 4(6)+(^-4)$$

or
$$5y < 20$$

i.e.
$$y < 4.$$

Is this consistent with the graph in Figure 8?

We can say, therefore, that in no circumstances can y be greater than 4. But there is no other possible combination of A and C that eliminates one or other of x and y, and so there is nothing else that we can say about x or y separately. This, of course, is what we would expect from the graph.

Exercise K

1 Solve the following graphically.

(a) $\begin{cases} x+y > 1 \\ x+2y < 2 \end{cases}$ (b) $\begin{cases} 2x+y > 3 \\ 2y > x+4 \end{cases}$

(c) $\begin{cases} 2x-y > 3 \\ x-2y < ^-4 \end{cases}$ (d) $\begin{cases} 2x+y > 3 \\ 2y+x > 4 \end{cases}$

2 Obtain algebraically the separate limits on x and y (where possible) for the first three parts of question **1**.

Summary

(1) Many simple equations in one unknown can be solved by the use of inverse operations. The flow diagram for the equation is constructed, and then the reversed flow diagram is obtained by reversing the order of the operations, and replacing each operation by its inverse. For example, the equation $\tfrac{1}{2}x-6 = 11$ can be represented by the flow diagram

$$x \longrightarrow \boxed{M\tfrac{1}{2}} \longrightarrow \boxed{S\,6} \longrightarrow 11$$

which when reversed is

$$11 \longrightarrow \boxed{A\,6} \xrightarrow{17} \boxed{D\tfrac{1}{2}} \longrightarrow 34 \quad (\boxed{D\tfrac{1}{2}} = \boxed{M\,2})$$

and so the solution is $x = 34$.

(2) An alternative method, using inverse elements, is to carry out the following steps.
   (a) 'Remove' (or multiply out) any brackets: $\frac{1}{2}x - 6 = 11$
   (b) If necessary, replace '$-n$' terms by '$+^-n$': $\frac{1}{2}x + ^-6 = 11$
   (c) Collect 'x on the left, and numbers on the right' by adding additive inverses to both sides: $\frac{1}{2}x = 11 + ^+6 = 17$
   (d) Obtain an expression for $1x$ by multiplying by the multiplicative inverse: $x = 2(17)$
   giving an answer $x = 34$.
(3) The solutions of equations and orderings should always be checked by substituting the answers in each side of the original relation, and checking that they do, in fact, balance.
(4) One-dimensional orderings can be solved in the same way as in (2), *but* it must be remembered that if you multiply both sides of an ordering by a *negative* number the direction of the ordering is changed. For example, if

$$^-3x < 24$$

then

$$x > ^-\tfrac{1}{3}(24).$$

(5) The solution set of simultaneous one-dimensional orderings is the intersection of the two separate solutions; i.e. if

$$A < B < C \quad \text{then} \quad A < B \quad \text{and} \quad B < C.$$

(6) An equation in two unknowns is represented by a line on an x, y graph. An ordering in two unknowns is a region of the graph, bounded by the line obtained by making an equality from the inequality.
(7) Simultaneous equations in two unknowns may be solved
   (a) graphically,
   (b) algebraically, using the substitution method,
   (c) algebraically, using the elimination method,
   (d) by the use of matrices.
(8) Simultaneous inequalities should be solved graphically. For each ordering it is usually better to shade *out* the unwanted region, and the solution set is then the area left unshaded at the end.

Post-test

Solve the following.

1 (a) $1 - 3x = 19$ (b) $2x + 5 = \frac{1}{2}(x - 5)$ (c) $\dfrac{x+5}{4} = 2 - 5x$

2 (a) $3x - 7 > 2$ (b) $4 \leqslant 1 - 2x$ (c) $2x < 3 < 5x + 6$

3 (a) $\begin{cases} 3x - y = 6 \\ x + 2y = 2 \end{cases}$ (b) $\begin{cases} 2x + 3y = 7 \\ 4y = 1 - x \end{cases}$ (c) $\begin{cases} y = x + 7 \\ 3y - 2x = 14 \end{cases}$

4 (a) $\begin{cases} 3x - y < 6 \\ x + 2y < 4 \end{cases}$ (b) $\begin{cases} 3x - y < 6 \\ x + 2y > 4 \end{cases}$

Assignment

Solve

1   $\frac{1}{2}(4-3x) = {}^-7$ by the method of inverse operations.

2   (a) $5x-6 = 12+9x$ using inverse elements
    (b) $2(3-x) = \frac{1}{3}(7+5x)$

3   (a) $2(1-x) < 3x+12$
    (b) $3x-7 < 2x-3 \leqslant x-1$

4   (a) $\begin{cases} 2x = 3y-6 \\ 4-x = y \end{cases}$ (b) $\begin{cases} 3x-7y = 8 \\ 5x+2y = 68 \end{cases}$ (c) $\begin{cases} x-y > {}^-2 \\ 3y+x < 2 \end{cases}$
    (by substitution)   (by elimination)   (graphically)

Answers

Pre-test

1   (a) $^-2$ (b) $^+7$ or just 7   (c) $\frac{19}{20}$ (d) $\frac{5}{6}$
    (e) $^+1$, or just 1   (f) $= \frac{3}{4} \times \frac{-6}{1} = {}^-4\frac{1}{2}$
2   (a) $^-3\frac{1}{2}$ (b) $3\frac{1}{2}$ (c) $2\frac{1}{2}$ (because $2x = 5$)   (d) $^+49$
3   (a) $2x+6$ (b) $\frac{1}{2}x-3$ (c) $-24+4x$
4   (a)

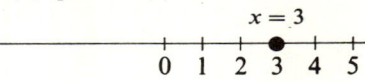

    (b)

5   See Figure A.

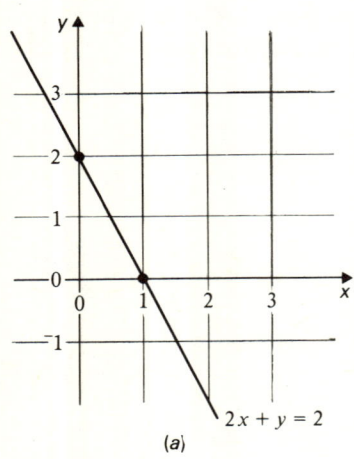
(a)

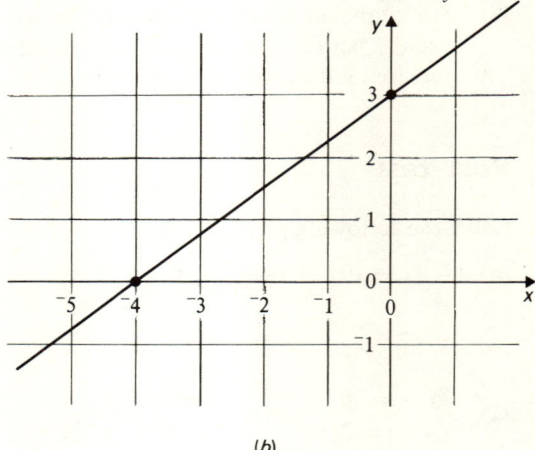
(b)

Figure A

20

1.1 Simple equations: solution by inverse operations

Using flow diagrams

> 1 $x \longrightarrow \boxed{S\ 7} \longrightarrow \boxed{D\ 5} \longrightarrow 1$
>
> 2 $4 \longrightarrow \boxed{A\ 6} \longrightarrow \boxed{D\ 4} \longrightarrow \boxed{S\ 4} \longrightarrow x$
>
> 3 $x \longrightarrow \boxed{D\ 5} \longrightarrow \boxed{S\ 7} \longrightarrow 1$
>
> 4 $4 \longrightarrow \boxed{S\ 4} \longrightarrow \boxed{D\ 4} \longrightarrow \boxed{A\ 6} \longrightarrow x$

In **2**, $x = -1\tfrac{1}{2}$, but in **4** $x = 6$. Working backwards, the example is true for $x = 4$, **1** is true for $x = 12$, and **3** is true for $x = 40$. By comparing the answers to **2** and **4**, and to **1** and **3**, we can see that the order of operations *does* matter.

> 5   Example **3** can be written

$$x \xrightarrow{\ \ x\ \ } \boxed{D\ 5} \xrightarrow{\ \ x/5\ \ } \boxed{S\ 7} \xrightarrow{\ \ x/5-7\ \ } 1$$

giving the relation $\dfrac{x}{5} - 7 = 1$.

Exercise A

> 1   (a) 37   (b) $-\tfrac{3}{4}$ (c) -2 (d) -1
>
> In (d), for example, the successive stages are
> $4 \div (-2) = -2,\ -2-(-3) = {}^+1,\ 1 \times (-1) = -1$

2   (a) $3(x+2) = 14$ (b) $\dfrac{x-2}{2} + 2 = 3$ (or $\tfrac{1}{2}(x-2)+2 = 3$)

     (c) $\tfrac{1}{2}(4-2x) = 3$ (or $\tfrac{1}{2}(-2x+4) = 3$)

     (d) $\dfrac{\tfrac{x}{3}+4}{2} = 1$ (or $\tfrac{1}{2}\left(\tfrac{x}{3}+4\right) = 1$)

> 3   (a) $x \longrightarrow \boxed{A\ 3} \longrightarrow \boxed{M\ 4} \longrightarrow 6$
>   (b) $x \longrightarrow \boxed{M\ 2} \longrightarrow \boxed{A\ 3} \longrightarrow 6$
>   (c) $x \longrightarrow \boxed{S\ 4} \longrightarrow \boxed{D\ 3} \longrightarrow 5$
>   (d) $x \longrightarrow \boxed{M\ 5} \longrightarrow \boxed{A\ 1} \longrightarrow \boxed{D\ 3} \longrightarrow 12$

Reversing the flow diagram

> 1   In each case the answer is x; for example $x+6-6 = x$.

> 2   The original flow diagram is $x \longrightarrow \boxed{D\ 5} \longrightarrow \boxed{S\ 7} \longrightarrow 1$
>
> which when reversed becomes $1 \xrightarrow{\ 1\ } \boxed{A\ 7} \xrightarrow{\ 8\ } \boxed{M\ 5} \xrightarrow{\ 40\ } 40$
>
> so that the solution to the equation
>
> $$\dfrac{x}{5} - 7 = 1 \text{ is } x = 40.$$

Exercise B

1 For question **2(a)** of Exercise A the reversed flow diagram is

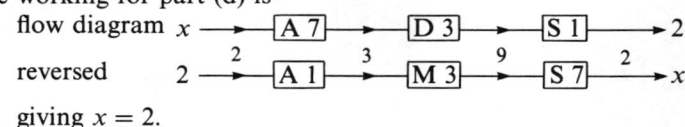

giving $x = 2\frac{2}{3}$.

Similarly, the solutions to the other parts of question **2** are
(b) $x = 4$ (c) $x = {}^-1$ (d) $x = {}^-6$
For question **3** the solutions are
(a) $-1\frac{1}{2}$ (b) $1\frac{1}{2}$ (c) 19 (d) 7

2 (a) $x = 5\frac{1}{2}$ (b) $x = {}^-4$ (c) $x = {}^-2$ (d) $x = 2$
The working for part (d) is

flow diagram $x \longrightarrow \boxed{A\ 7} \longrightarrow \boxed{D\ 3} \longrightarrow \boxed{S\ 1} \longrightarrow 2$

reversed $\quad 2 \xrightarrow{2} \boxed{A\ 1} \xrightarrow{3} \boxed{M\ 3} \xrightarrow{9} \boxed{S\ 7} \xrightarrow{2} x$

giving $x = 2$.

Self-inverse operations

1 (a) $f(3) = 4 - 3 = 1$ (b) $ff(3) = f(1) = 4 - 1 = 3$
(c) $g(3) = 12/3 = 4$ (d) $gg(3) = g(4) = 12/4 = 3$
2 Similarly (a) $ff(5) = 5$ (b) $ff(^-2) = {}^-2$ (c) $gg(5) = 5$ (d) $gg(^-2) = {}^-2$

Exercise C

1	$x = 4\frac{1}{2}$	**7**	$x = 2$
2	$x = {}^-1\frac{1}{2}$	**8**	$x = {}^-2$
3	$x = 3\frac{1}{2}$	**9**	$x = 16$
4	$x = 10$	**10**	$x = {}^-12$
5	$x = 14$	**11**	$x = 0$
6	$x = 18$	**12**	$x = 12$

For example:

4 The flow diagram is

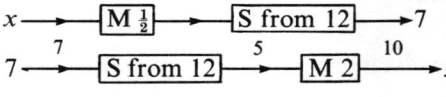

reversed is

giving $x = 10$

8 The flow diagram is

reversed is

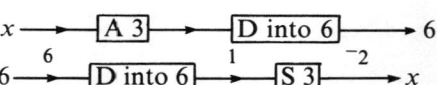

giving $x = {}^-2$

12 The flow diagram is

reversed is

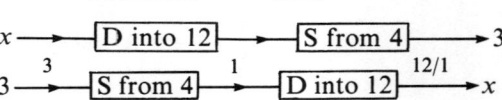

giving $x = 12$

1.2 The use of inverse elements

Exercise D

> 1 $\boxed{A\ ^-5}$ $7x+5+^-5 = 8+^-5$ 2   rewrite as $5x+^-6 = 14$
                $7x = 3$ $\boxed{A\ ^+6}$ $5x = 20$
     $\boxed{M\ \tfrac{1}{7}}$ $\tfrac{1}{7} \times 7x = \tfrac{1}{7} \times 3$ $\boxed{M\ \tfrac{1}{5}}$ $x = 4$
     or $x = \tfrac{3}{7}$

3   $\boxed{A\ ^-2}$ $2-3x+^-2 = 26+^-2$ *or*   rewrite as $2+^-3x = 26$
                    $^-3x = 24$ $\boxed{A\ ^+3x}$ $2 = 26+3x$
     $\boxed{M\ ^-\tfrac{1}{3}}$ $x = ^-8$ $\boxed{A\ ^-26}$ $^-24 = 3x$
                                                   $\boxed{M\ \tfrac{1}{3}}$ $^-8 = x$
                                                   or $x = ^-8$

4   $\boxed{M\ \tfrac{1}{3}}$ $(2x-1) = \tfrac{8}{3} = 2\tfrac{2}{3}$
     $\boxed{A\ ^+1}$ $2x = 3\tfrac{2}{3}$
     $\boxed{M\ \tfrac{1}{2}}$ $x = \tfrac{11}{3} \times \tfrac{1}{2} = \tfrac{11}{6} = 1\tfrac{5}{6}$
(But see also the section on 'Using the distributive law'.)

5   $\boxed{A\ ^-\tfrac{1}{4}}$, $\boxed{M\ \tfrac{8}{3}}$, giving $x = ^-\tfrac{1}{3}$.

6   $\boxed{M\ 3}$, $\boxed{A\ ^-2}$, giving $x = 10$.

Compare the two ways shown for question **3** with the method using the inverse operations (one self-inverse in this case).

     The flow diagram is $x \longrightarrow \boxed{M\ 3} \longrightarrow \boxed{\text{S from 2}} \longrightarrow 26$
                                             26 $^-24$ $^-\tfrac{24}{3}$
     which reversed gives $26 \longrightarrow \boxed{\text{S from 2}} \longrightarrow \boxed{D\ 3} \longrightarrow x$
     or $x = ^-8$

Note that, for example, $\boxed{M\ \tfrac{1}{2}}$ is the same as $\boxed{D\ 2}$; to find half of something you divide by 2. Both are the inverse of $\boxed{M\ 2}$, the first uses the inverse element, the second the inverse operation.

Exercise E

> 1   (a) $\tfrac{13}{42}x$ (b) $\tfrac{1}{4}(x+y)$ (c) $15-6x$ (d) $-12+8x$

> 2   (a) $x = 1$ (b) $x = \tfrac{1}{6}$ (c) $x = 11$ (d) $x = ^-1$
     (e) $x = ^-1\tfrac{3}{4}$ (f) $x = \tfrac{7}{3}$ (g) $x = 1\tfrac{1}{6}$ (h) $x = ^-5$
The working for **2(f)** would be

     $\boxed{M\ 4}$ $12 - 2(x+1) = (x-1) + 4$

(Note the brackets round $x+1$. Remember that $\dfrac{x+1}{2}$ is the same as $\tfrac{1}{2}(x+1)$.)

remove brackets $12 - 2x - 2 = x - 1 + 4$
or $10 - 2x = x + 3$
$\boxed{A\ ^-x}$
$\boxed{A\ ^-10}$ $10 - 3x = 3$
                       $^-3x = ^-7$
$\boxed{M\ ^-\tfrac{1}{3}}$ $x = ^+\tfrac{7}{3}$ (or $2\tfrac{1}{3}$)

1.3 One-dimensional orderings

1 LHS becomes $3+7 = 10$; RHS becomes $5+7 = 12$; and $10 < 12$.
2 LHS $= 3+{}^-6 = {}^-3$; RHS $= 5+{}^-6 = {}^-1$; and ${}^-6 < {}^-1$.

Similarly in **3, 4** and **5**, the left-hand side remains 'lighter' than (i.e. less than) the right-hand side.

Exercise F

1	$x > {}^-6$	**5**	$x \leqslant {}^-2$
2	$x < {}^-5$	**6**	$x < 2$
3	$x > {}^-1\frac{1}{4}$	**7**	$x > {}^-3$
4	$x \leqslant 3$	**8**	$x \leqslant 7$

Exercise G

1 (a) [number line from -4 to 4]   (b) [number line from -5 to 1]

2 No. It is not possible for x to be greater than 0 *and* less than $^-4$ at the same time.
3 (a) $3x > x+2$ simplifies to $x > 1$
   $x+2 > 2x-4$ simplifies to $x < 6$
the simultaneous solution is $1 < x < 6$.
(b) $x < {}^-3$ and $x > \frac{1}{2}$. No solution.
(c) $x > {}^-3$ and $x \leqslant \frac{1}{2}$, i.e. ${}^-3 < x \leqslant \frac{1}{2}$.
(d) $x > 1$ and $x > 3$. The intersection of these is $x > 3$.

1.4 Equations in two unknowns

Exercise H

1 Solution is $x = 2$, $y = 4$ (see Figure B).

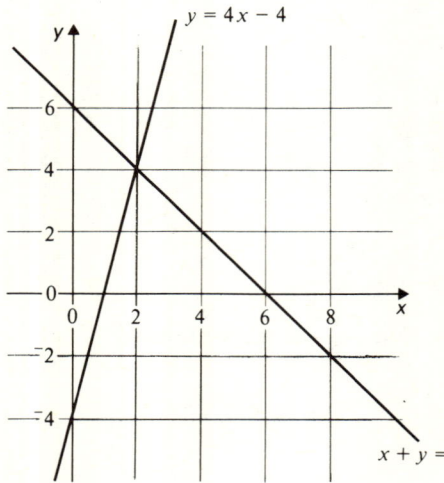

Figure B

2 Solution is $x = -2\frac{1}{2}$, $y = -\frac{1}{2}$ (see Figure C).

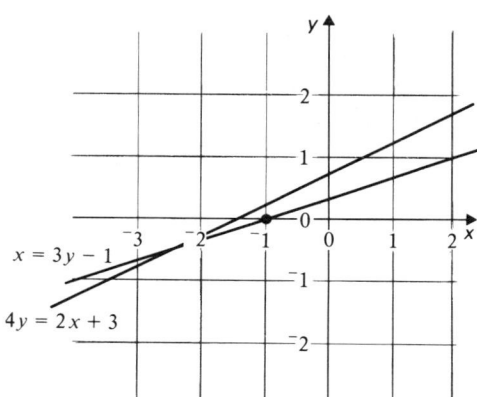

Figure C

3 Solution is $x = -\frac{1}{2}$, $y = 2$ (see Figure D).

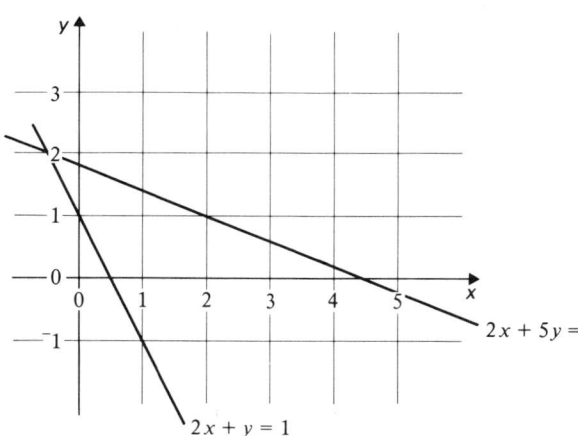

Figure D

Algebraic methods

> **1** (a) Substitute $\frac{1}{4}(2x+3)$ for y in (i) $x = 3 \times \frac{1}{4}(2x+3) - 1$
         $\boxed{M\ 4}$ $4x = 3(2x+3) - 4$
         remove brackets $4x = 6x + 9 - 4$
         $\boxed{A\ ^-6x}$ $^-2x = 5$
         $\boxed{M\ ^-\frac{1}{2}}$ $x = -2\frac{1}{2}$
         substitute in $y = \frac{1}{4}(2x+3)$ $y = \frac{1}{4}(^-5+3) = -\frac{1}{2}$

(b) Substitute $(3y-1)$ for x in (ii)
      remove brackets

$$4y = 2(3y-1)+3$$
$$4y = 6y-2+3$$
$$^-2y = 1$$

A ^-6y
M $^-\frac{1}{2}$

$$y = ^-\tfrac{1}{2}$$

Substitute in $x = 3y-1$ $x = 3(^-\tfrac{1}{2})-1 = ^-2\tfrac{1}{2}$

(c) The second method is probably easier as it involves less work with fractions.

Exercise J

1. (a) Substitute $(4+y)$ for x in the first equation
$$(4+y)+2y = 10$$
$$3y = 6$$
$$y = 2 \quad \text{and} \quad x = 4+2 = 6$$
(Alternative substitutions are $(x-4)$ for y in the first equation, or $(10-2y)$ for x in the second equation.)

(b) Substitute $(3x-1)$ for y in the first equation
$$2x+3(3x-1) = 1$$
$$2x+9x-3 = 1$$
$$x = \tfrac{4}{11} \quad \text{and} \quad y = \tfrac{1}{11}$$

2. (a) Adding the two lines $2x = 11$
                     $x = 5\tfrac{1}{2}$
  Subtracting $2y = 5$
                     $y = 2\tfrac{1}{2}$

(Alternatively, y can be found by substituting $5\tfrac{1}{2}$ for x in the first equation.)

(b) first equation -4(second equation) gives
$$4x-3y-4(x-2y) = 1-4(4)$$
or $4x-3y-4x+8y = 1-16$
$$5y = ^-15$$
$$y = ^-3$$
From the second equation $x = 4+2y$,
hence
$$x = ^-2$$

(c) 3(first equation) $+$ 2(second equation) gives $13x = 0$, i.e. $x = 0$.
Substitution, or 2(first) $-$ 3(second), gives $y = ^-2$.

1.5 Orderings in two unknowns

Simultaneous orderings

1. (a) $3 > 0$ $3 > 0$ $6 > 2$ $6 > 2$
       $2 > 1$ $5 > 1$ $3 < 5$ $3 < 9$
       $\overline{5 > 1}$ $\overline{8 > 1}$ $\overline{9 \;?\; 7}$ $\overline{9 \;?\; 11}$

If the directions of the inequalities are not the same, it is not possible to say whether the final result is $>$ or $<$.

(b) $3 > 0$ $3 > 0$ $6 > 2$ $6 > 2$
       $2 > 1$ $5 > 1$ $3 < 5$ $3 < 9$
       $\overline{1 \;?\; ^-1}$ $\overline{^-2 \;?\; ^-1}$ $\overline{3 > ^-3}$ $\overline{3 > ^-7}$

The only generalisation that we can make is that
if $a > b$
and $c < d$
then $a - c > b - d$.

Exercise K

1 See Figure E, parts (a)–(d).

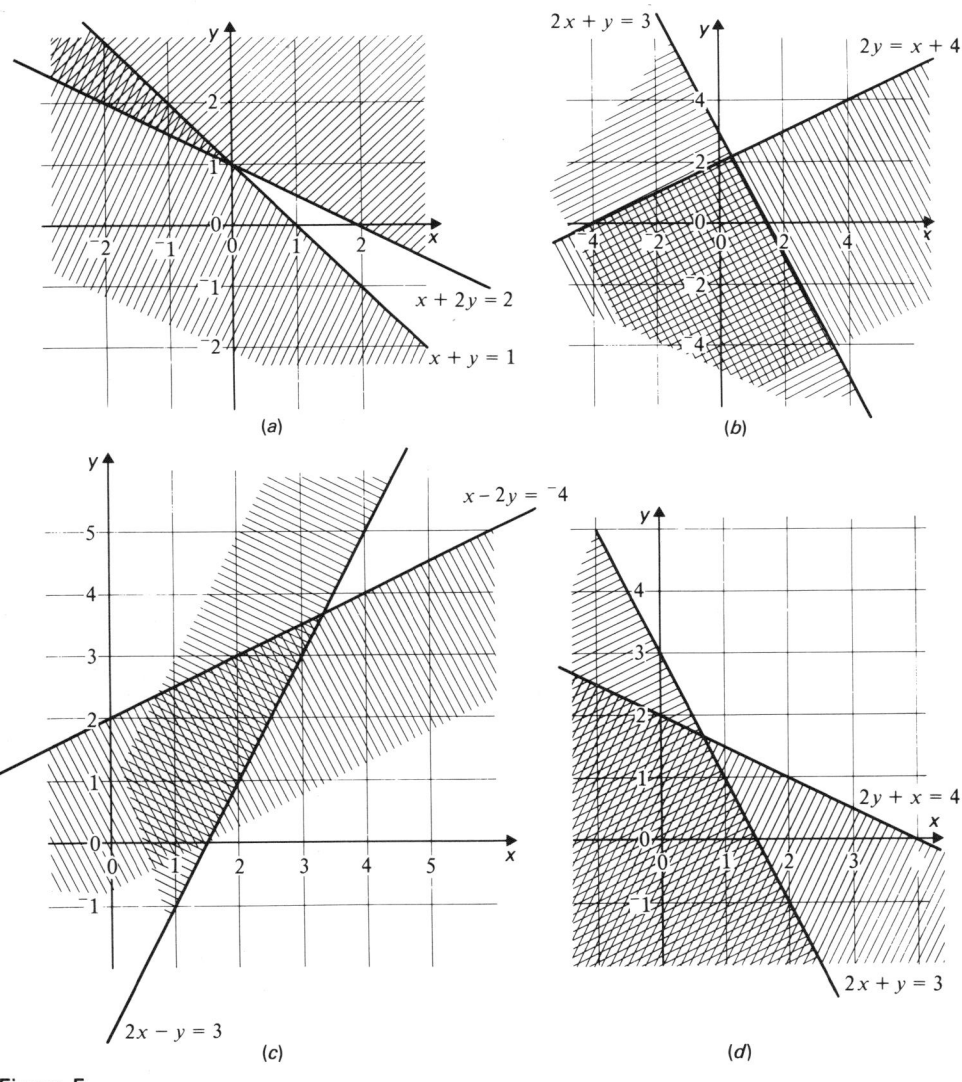

Figure E

2 (a) Using the 'combinations' $2A - C > 2B - D$ gives $x > 0$, and $A - C > B - D$ gives $^-y > ^-1$ or $y < 1$.

(b) Rewriting the second ordering as $^-x+2y > 4$, and using the combination $A+2C > B+2D$ gives $y > \frac{11}{5}$.

(c) The combination $A-2C > B-2D$ gives $3y > 11$, and $2A-C > 2B-D$ gives $3x > 10$, i.e. $x > 3\frac{1}{3}$ and $y > 3\frac{2}{3}$ separately.

Post-test

1. (a) $x = {}^-6$ (b) $x = {}^-5$ (c) $x = \frac{1}{7}$
2. (a) $x > 3$ (b) $x \leqslant {}^-1\frac{1}{2}$ (c) $x < 1\frac{1}{2}$ and $x > {}^-\frac{3}{5}$, i.e. ${}^-\frac{3}{5} < x < 1\frac{1}{2}$
3. (a) $x = 2, y = 0$
   Elimination (twice the first + second) is a suitable start.
   (b) $x = 5, y = {}^-1$
   Substitution $(1-4y$ for $x)$ in the first is a possible start, or elimination (first − twice the second) after the second equation is rewritten as $x+4y = 1$.
   (c) $x = {}^-7, y = 0$
   Substitution of $x+7$ for y in the second equation is the easiest start.
   All of these, of course, could be solved graphically.
4. See Figure F.

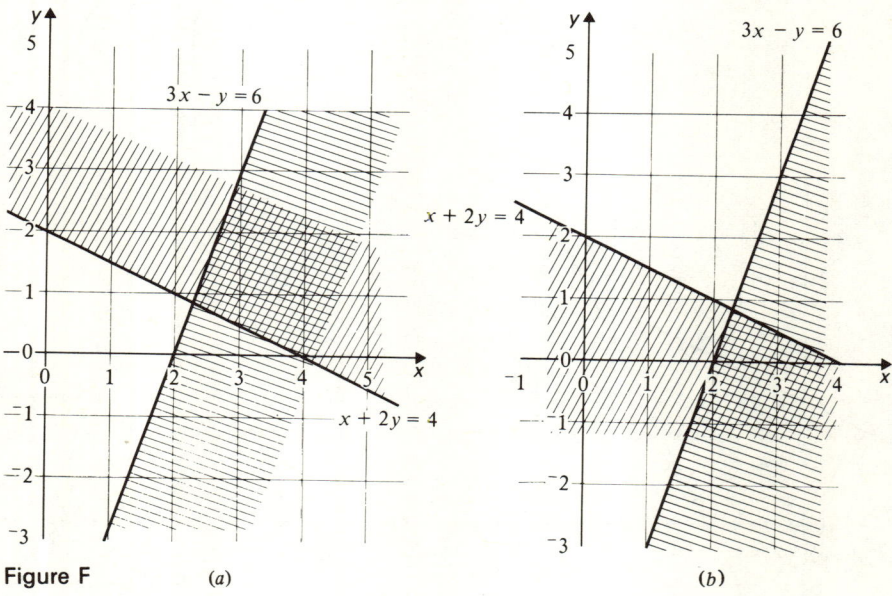

Figure F   (a)   (b)

2 Formulas

Objectives

This is what you should be able to do after studying this chapter.
(1) Construct a formula when you are given the necessary information.
(2) Use a formula to calculate the value of one of the variables when the values of all the other variables are given.
(3) Change the subject of a formula.

Pre-test

> 1 (a) Construct the flow charts for the following expressions, starting with x in each case: $3(1+2x)$, $b+\frac{1}{2}cx^2$, $2\sqrt{(4+x)}$
> (b) Write down the reversed flow diagrams for the expressions in part (a), and hence find x in the following equations: $3(1+2x) = 4$, $a = b+\frac{1}{2}cx^2$, $2\sqrt{(4+x)} = {}^-7$

2 Write down the values of a for each of the following equations. (In most cases you shouldn't need to do any work on paper.)
(a) $2a = 3$ (b) $a+2 = 3$ (c) $2a-3 = 5$
(d) $a^2 = 4$ (e) $a^3 = 28$ (correct to the nearest integer) (f) $\sqrt{a} = 2.4$

3 Write down the formulas for the following
(a) The circumference (C cm) of a circle of radius r cm
(b) The area (A cm²) of a circle of radius r cm
(c) The perimeter (p m) of a square of side x m
(d) The perimeter (p cm) of a rectangle of sides x cm and y cm

2.1 Construction of formulas

$$T = 2\pi\sqrt{\frac{L}{g}}, \quad \frac{1}{u}+\frac{1}{v}=\frac{1}{f}, \quad E = mc^2, \quad v^2 = u^2+2fs$$

These are all well-known formulas used in science. A *formula* expresses a relation between two or more variables associated with a particular object or situation. Thus the first formula above states a relation between three variables: the length (L) of a pendulum, the time (T) for one swing of the pendulum, and the acceleration due to gravity (g) at the place where the pendulum is.

If we 'read' the formula as a sentence, it starts 'T is equal to...', and so we describe T as 'the subject of the formula'.

We might well ask how such formulas were obtained. Initially some were deduced by carrying out experiments and noticing that the results satisfied a certain relationship (see Example I below), but eventually most are obtained from definitions and theoretical considerations. Example II below is a compromise between these methods.

Example I: The table below shows some figures given in the *Highway Code* for the stopping distances, in metres, for cars travelling at various speeds, in kilometres per hour, on a dry road.

Speed (kph)	40	60	80	100
Stopping distance (m)	16	36	64	100

Let the speed be V kph, and the corresponding stopping distance be D m. If we can find a relation connecting V and D that is satisfied by all the pairs of values in the table, we shall have found a formula connecting V and D.

One way of seeing whether there is a relation between two sets of figures such as these is to plot them on a graph (see Figure 1).

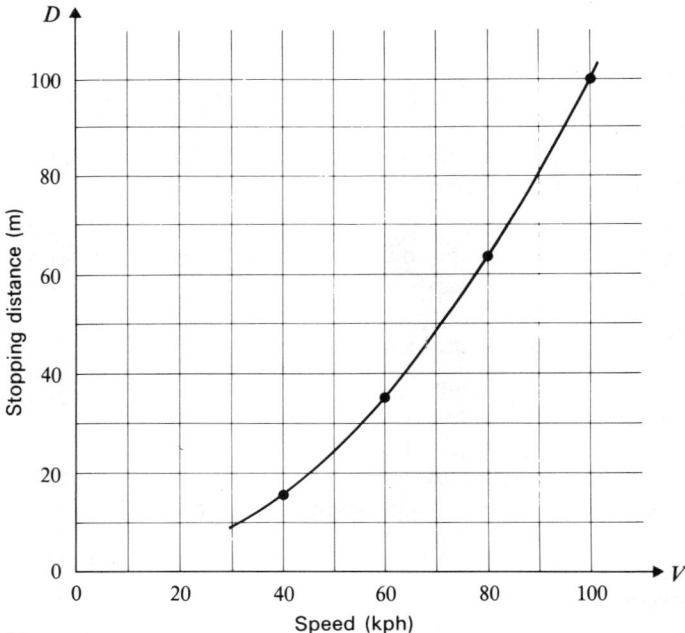

Figure 1

1. What is D when $V = 0$? If this point is added to the graph, and the curve drawn through it, can you recognise the curve? This should lead you to expect that a simple relation exists between D and V^2. Complete this table

D	16	36	64	100
V^2	1600			

30

and write down the relation between D and V^2. Use your result to find the stopping distances for cars travelling at (a) 50 kph (b) 75 kph.

Example II: It is noted that the volume of a soap bubble increases at the rate of 7 cm³/s, and initially the volume is 2 cm³. If, after t s, the volume is V cm³, obtain a formula for V in terms of t (i.e. a formula with V as the subject, and for which the right-hand side involves the variable t as well as numbers).

After 1 s the volume has *increased* by 7 cm³, so that the total volume is $(2+7)$ cm³.
After 2 s the volume has increased by another 7 cm³, so that the total volume is now $(2+7\times 2)$ cm³.
After t s the volume has increased by a total of $7\times t$ cm³, so that the total volume is $(2+7\times t)$ cm³.
Hence the required formula is $V = 2+7t$.

Exercise A

1. The roasting time for a joint of meat is 45 min for each kilogram, plus 45 min extra. If a joint of mass M kg takes T min altogether, find the formula for T in terms of M.
2. A square garden of side L m consists of a square lawn of side s m surrounded by a path of width d m. (The lawn and the path between them occupy all of the garden.)
   (a) Draw a diagram to illustrate this.
   (b) If the area of the garden is A m², write down the formula connecting A and L.
   (c) Write down a relation connecting s, L and d.
   (d) If the area of the lawn is B m², write down a formula for B in terms of L and d.
   (e) The cost of concreting the path is £ t per square metre. If the total cost is £ C, find a formula connecting C, L, s and t.
3. A fruiterer buys a box of 6 pineapples for 60 p. He finds that x of them are bad, and he sells the others at 18 p each. If his profit is P p, find a formula for P in terms of x, and hence the greatest number that can be bad before he ceases to make a profit.

2.2 Calculations using formulas

A formula is *used* when we find the numerical value of one of the variables having been given the values of all the others.

For example, we can use the (approximate) formula for the surface area of a cylindrical tin
$$A = 6r(r+h)$$
to find the surface area (A) when we know the values of the radius (r) and the height (h).

By the nature of its construction, it is easiest to use a formula to find the value of the subject: the formula above is designed to tell you what the value of A is when you are given the values of r and h. It is a little more difficult to find the value of, say, h when given the values of A and r; we now have an equation to solve.

1. Using the formula above, find the following
   (a) The value of A when $r = 3$ and $h = 5$
   (b) The value of h when $r = 4$ and $A = 264$
   (c) The value of r when $h = 6$ and $A = 330$ (if you can!)

A little common sense is necessary when considering the problem of units. In a formula such as $T = 2\pi\sqrt{(L/g)}$, *physicists* would say that if $L = 50$ cm and $g = 980$ cm/s², then T is 1.42 s. *Mathematicians* usually prefer a letter to stand for just a number, and would say that this formula is for the swing-time (T *seconds*) of a pendulum of length L *cm* at a spot where the gravitational acceleration is g *cm/s²*, and would therefore say that when $g = 980$ and $L = 50$ then $T = 1.42$.

The main thing is to be consistent in sticking to either the physicists' approach or the mathematicians' approach, and, if using the physicists' approach, to be consistent within the formula, e.g. don't give acceleration in *cm/s²* and the length in *metres*. (Notice, therefore, that the formula derived in Example I of the section 'Construction of formulas' is a mathematicians' formula!).

Exercise B

1. $m = \frac{1}{2}(x+y)$
   (a) Find the value of m when $x = 6$ and $y = -4$.
   (b) Find the value of x when $m = -4$ and $y = 2$.

2. $A = 6 - \dfrac{12}{r}$
   (a) Find the value of A when $r = 8$.
   (b) Find the value of r when $A = 2$.

3. The maximum safe speed for a cyclist riding round a circular track of radius r m is V kph where
   $$V = \tfrac{12}{5}\sqrt{(20r)}.$$
   (a) Find the maximum safe speed on a track of radius 5 m.
   (b) Find the radius of the track for which the maximum safe speed is 36 kph.

4. $F = \dfrac{mv^2}{r}$
   (a) Find F when $m = 12$, $v = 3$ and $r = 8$.
   (b) Find v when $F = 48$, $m = 4$ and $r = 3$.

5. $T = \dfrac{12}{\sqrt{c}}$
   (a) Find T when $c = 64$.
   (b) Find c when $T = 4$.

2.3 Changing the subject of a formula

1. Write down the subject of each formula in Exercise B.

In the second part of each question in Exercise B you were asked to find the value of a variable that was not the subject of the formula. For example in question 3(b), putting $V = 36$ we have to find r from the statement
$$36 = \tfrac{12}{5}\sqrt{(20r)}.$$

In other words, we had to solve the equation $\frac{12}{5}\sqrt{(20r)} = 36$, and you may have done this by writing down the flow diagram

$$r \longrightarrow \boxed{M\ 20} \longrightarrow \boxed{\text{take }\sqrt{}} \longrightarrow \boxed{M\ \tfrac{12}{5}} \longrightarrow 36$$

and reversing this to give

$$36 \longrightarrow \boxed{M\ \tfrac{5}{12}} \longrightarrow \boxed{\text{square}} \longrightarrow \boxed{D\ 20} \longrightarrow 11.25$$

so that $r = 11.25$.

When a formula is often used to find the value of a variable that is not the subject, it becomes worth while to rearrange it so that the required variable *is* the subject. This process is called 'changing the subject of a formula' and is carried out as though we were solving an equation.

The 'unknown' is the variable that we want to isolate (i.e. make the subject), and all the other variables are to be treated as known numbers. The only difference is that we cannot simplify or evaluate such expressions as $r+3$ or $5r$. (In written work it can be helpful to emphasise which variable is the 'unknown' by underlining it, ringing it, or writing it in a different colour.)

Example 1: To make r the subject of the formula $V = \frac{12}{5}\sqrt{(20r)}$.

Write the formula as $\quad \frac{12}{5}\sqrt{(20r)} = V$

$\boxed{M\ \tfrac{5}{12}} \qquad \sqrt{(20r)} = \dfrac{5V}{12}$

$\boxed{\text{square}} \qquad 20r = \dfrac{25V^2}{144}$

$\boxed{M\ \tfrac{1}{20}} \qquad r = \dfrac{5V^2}{576}$

Example 2: To make r the subject of the formula $A = 6 - \dfrac{12}{r}$.

The flow diagram for A is

$$r \longrightarrow \boxed{D\ \text{into}\ 12} \longrightarrow \boxed{S\ \text{from}\ 6} \longrightarrow A$$

reversing this, we have

$$A \longrightarrow \boxed{S\ \text{from}\ 6} \longrightarrow \boxed{D\ \text{into}\ 12} \longrightarrow r$$

so that the rearranged formula is $\quad r = \dfrac{12}{6-A}$.

Exercise C

1. Rearrange the following formulas so that the letter in the square bracket becomes the subject.
   (a) $A = 3b$ $\quad [b]$ $\qquad$ (b) $V = IR$ $\quad [I]$ $\qquad$ (c) $v = u + ft$ $\quad [f]$
   (d) $m = \tfrac{1}{2}(x+y)$ $\quad [x]$ $\qquad$ (e) $F = \dfrac{mv^2}{r}$ $\quad [v]$ $\qquad$ (f) $F = \dfrac{mv^2}{r}$ $\quad [r]$
   (g) $v^2 = u^2 + 2as$ $\quad [u]$ $\qquad$ (h) $s = \dfrac{(u+v)}{2} t$ $\quad [v]$

2. Make L the subject of the formula $T = 2\pi\sqrt{(L/g)}$, and hence find the value of L when $T = 1.2$ and $g = 10$. (Assume $\pi^2 = 10$.)

3 The total surface area, S cm², of a cylinder of radius r cm and height h cm, is given by the formula
$$S = 2\pi rh + 2\pi r^2.$$
Make h the subject of this formula and find the height of a cylinder that has a surface area of 84 cm² and a base radius of 2 cm.

4 A circle of radius r cm is cut from a larger circle of radius R cm.
  (a) If the area of the part remaining in the larger circle is A cm², find a formula for A in terms of R and r.
  (b) Make R the subject of this formula.
  (c) Find A when $r = 2.7$ and $R = 4.1$.
  (d) Find R when $A = 2$ and $r = 0.8$.

5 Use the formula $E = mc^2$ to find the value of c when $m = 0.03$ and $E = 2.7 \times 10^{19}$. Give the answer in standard form.

2.4 Further examples

When solving equations in Chapter 1 we found that there were times when the distributive law was used in one form or another, and this is also true when changing the subject of a formula. Consider the formula
$$x = ab + 2b$$
This can be rewritten in the form $b = \frac{1}{2}(x - ab)$, but is it true to say that b is now the subject of this formula? As there is still a term involving b on the right-hand side, clearly we cannot evaluate b easily from this 'rearranged' formula.

But, returning to the original formula, we notice that b occurs in both terms on the right-hand side, and we apply the distributive law
$$x = b(a+2) \quad \text{or} \quad x = (a+2)b$$
which we rewrite as
$$(a+2)b = x$$
Multiplying both sides by $\dfrac{1}{a+2}$ gives
$$b = \frac{x}{a+2}$$
as the rearranged formula with b as the subject.

1 In which of these three formulas is x the subject?
  (a) $x = \dfrac{ax+by}{c}$ (b) $x = 3y - 7$ (c) $x = 3y^3 - 5x$
  Make x the subject of the other two.

2 Complete the following to make P the subject of $Q + Pt = Ps$.
$Q = \ldots$
$Q = P(\ldots)$
$P = \ldots$

A useful general procedure to follow in changing the subject (and in solving similar equations) is: 'remove' fractions; 'remove' brackets and/or square roots;

collect x-terms on the left-hand side; factorise the left-hand side; find what one 'x' is equal to.

The following example illustrates this.

Make x the subject of the formula $y = \sqrt{\left(\dfrac{5-x}{x}\right)}$.

boxed: square	$y^2 = \dfrac{5-x}{x}$	(removes $\sqrt{\ }$)
boxed: M x	$xy^2 = 5-x$	(removes fractions)
boxed: A ^+x	$xy^2 + x = 5$	(collects x on the LHS)
Use distributive law	$(y^2+1)x = 5$	(factorises LHS)
boxed: M $\dfrac{1}{y^2+1}$	$x = \dfrac{5}{y^2+1}$	(finds 'one x')

Exercise D

Rearrange the following formulas so that the letter in the square bracket becomes the subject.

1. $a = 3p + pq$ [p]
2. $s = 2ac + 4ab$ [a]
3. $y = (x+5)/x$ [x]
4. $p = \tfrac{1}{2}mv^2 - \tfrac{1}{2}mu^2$ [m]
5. $A = P + \dfrac{PRT}{100}$ [P]
6. $z = \dfrac{x}{x+2}$ [x]
7. $v = \dfrac{fu}{u-f}$ [f]
8. $w = \dfrac{2t-3}{3t-2}$ [t]

Summary

(1) A formula is a relation connecting two or more variable quantities.

$$C = 20000\left(k + \dfrac{12}{n}\right)$$

is a formula connecting the variables C, k and n. C is called the subject of the formula.

(2) Formulas are used to calculate the value of one of the variables when the values of all the others are known. For example, if we are told that $a = 2$ and $b = \tfrac{1}{2}$, we can find the value of G in

$$\dfrac{1}{G} = \dfrac{1}{a} + \dfrac{1}{b}$$

When the values for a and b are substituted, the formula becomes a simple equation, and we can calculate the value of G.

(3) When a formula is given, its form can be changed so that a different variable becomes the subject. This process is known as 'changing the subject of a formula'. The method is similar to that used in solving equations, and may be summarised by: 'clear' square roots, fractions and brackets; collect terms in 'x' on the left-hand side; factorise; and find the value of 1x.

Post-test

1. From the flow diagram below, write down the formula for the size of the interior angle ($a°$) of a regular polygon with n sides.

   $n \longrightarrow \boxed{\text{D into } 360} \longrightarrow \boxed{\text{S from } 180} \longrightarrow a$

   Find (a) a when $n = 15$ (b) n when $a = 140$.

2. Derive a formula for the base angle ($x°$) of an isosceles triangle as shown in Figure 2 in terms of the angle ($v°$) at the vertex. Find x when $v = 36$.

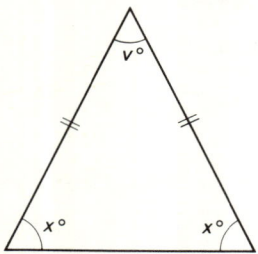

Figure 2

3. Use the formula $S = \pi r L$ to find (a) S when $r = 1.4$ and $L = 20$ (b) L when $S = 50$ and $r = 3.2$. (Give your answers to 3 figure accuracy, using either 3.14 or 22/7 for π.)

4. If $v = n\sqrt{(a^2 + x^2)}$ find n when $v = 72$, $a = 13$ and $x = 5$.

5. A passenger travelling by air is allowed to take 20 kg of luggage free. For any luggage over that amount the charge is $7 per kilogram. If the total charge for W kg is $C (provided $W > 20$), which of the following formulas is correct?
   (a) $C = 7W - 20$ (b) $C = 7(W - 20)$ (c) $C = 20 - 7W$ (d) $C = 7(W + 20)$

Assignment

1. Make x the subject of each of the following formulas.
   (a) $y = ax + b$ (b) $y = p(x + q)$ (c) $y + 5x = mx$
   (d) $y = \dfrac{x + k}{2x}$ (e) $y = \dfrac{x}{x - c}$ (f) $y = \dfrac{a + x}{a - x}$

2. To convert a temperature given in °F to the Réaumur scale (°R) the instructions are: (start F); subtract 32; multiply by 4; divide by 9.
   (a) Express this as a flow diagram, and as a formula.
   (b) Make F the subject of the formula.
   (c) What are the equivalent temperatures to 0 °F, 212 °F, 100 °R?

3. W, a and x are positive numbers such that $W^2 = a^2 - a^2 x^2$.
   (a) Find the value of W when $a = 5$ and $x = 0.6$.
   (b) Make x the subject of the formula, and hence find the value of x when $W = 9$ and $a = 10$.
   (c) Make a the subject of the formula, and find the value of a when $W = 9$ and $x = 0.8$.

36

4   Ohm's Law states that if a current of I amps flows through a resistance of R ohms, then the potential difference, V volts, across the resistance is given by $V = IR$.
   (a) If a current of 5 amps flows through a resistance of 480 ohms, calculate the potential difference.
   (b) Make I the subject of the formula.
   (c) The power consumed, W watts, is given by the formula $W = IV$. Combine this with the Ohm's Law formula to find W in terms of I and R, and in terms of V and R.
   (d) An electric fire consumes 2000 watts and the voltage (potential difference across the fire) of the supply is 250 volts. Find the resistance of the fire, and the current consumed.

5   The cost of building a concrete bridge depends on the number of spans used in the design. For a bridge with n spans, the cost, £ C, is given by the formula

$$C = 40\,000 \left(\frac{12}{n} + n\right)$$

Calculate the costs of bridges with 1, 2, 3, 4, 5 and 6 spans. Which is the most economical?

Answers

Pre-test

1 (a)

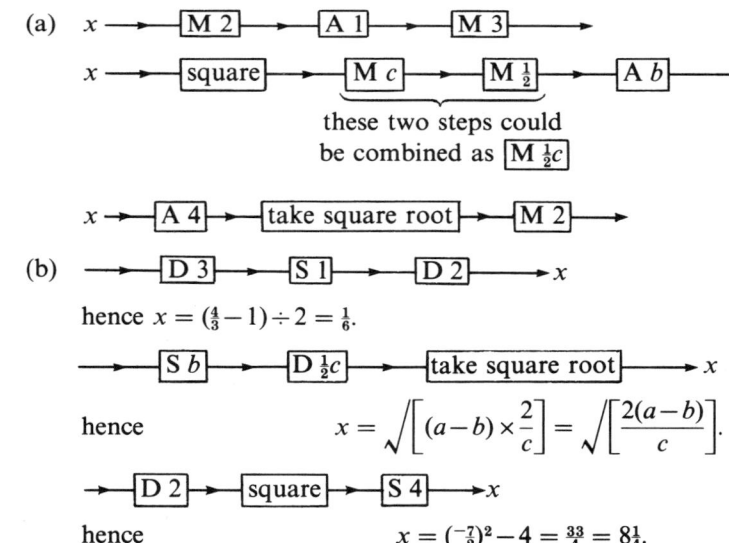

hence $x = (\frac{4}{3} - 1) \div 2 = \frac{1}{6}$.

hence $x = \sqrt{\left[(a-b) \times \frac{2}{c}\right]} = \sqrt{\left[\frac{2(a-b)}{c}\right]}$.

hence $x = (-\frac{7}{2})^2 - 4 = \frac{33}{4} = 8\frac{1}{4}$.

(Note, that in checking in the original equation it is necessary to take the negative square root of $12\frac{1}{4}$.)

2 (a) $1\frac{1}{2}$ (b) 1 (c) 4 (d) $^+2$ or $^-2$ (e) 3 (f) 5.76
3 (a) $C = 2\pi r$ (b) $A = \pi r^2$ (c) $p = 4x$ (d) $p = 2x + 2y$ or $p = 2(x+y)$

37

2.1 Construction of formulas

1 When $V = 0$, $D = 0$ as well. The curve looks like a parabola (i.e. of the form $y = kx^2$ for some value of k).

D	16	36	64	100
V^2	1600	3600	6400	10000

The relation is $D = V^2/100$.
(a) When $V = 50$, $D = 25$ and (b) when $V = 75$, $D = 56.25$ (≈ 56).

Exercise A

1 $T = 45M + 45$ or $T = 45(M+1)$

2 (a) See Figure A

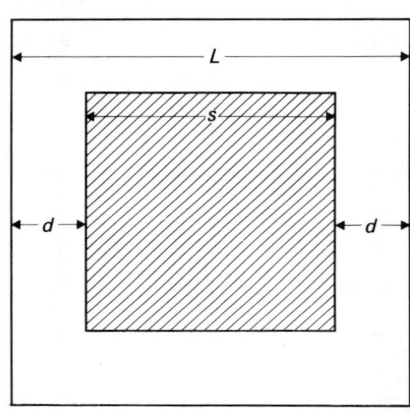

Figure A

(b) $A = L^2$ \hspace{2cm} (d) $B = (L-2d)^2$
(c) $s + 2d = L$ \hspace{1.5cm} (e) $C = (L^2 - s^2)t$
or $s = L - 2d$
or $d = \dfrac{L-s}{2}$

3 $P = 18(6-x) - 60$ or $P = 48 - 18x$
\hspace{3cm} If $P > 0$
\hspace{2cm} then $48 - 18x > 0$
\hspace{3cm} i.e. $x < \tfrac{8}{3}$
He makes a profit if not more than 2 pineapples are bad.

2.2 Calculations using formulas

1 (a) $A = 6 \times 3 \times (3+5) = 18 \times 8 = 144$
(b) Substituting in the formula gives $264 = 6 \times 4 \times (4+h)$
\hspace{4cm} or $264 = 24(4+h)$
\hspace{4cm} or $264 = 96 + 24h$

Hence $h = (264-96) \div 24 = 7$
(c) Substitution in the formula gives $330 = 6r(r+6)$

$$\text{or} \quad r(r+6) = 55$$

Since $5 \times 11 = 55$, a 'trial and error' answer is $r = 5$.

Exercise B

1. (a) $m = 1$
   (b) from $^-4 = \frac{1}{2}(x+2)$
   $x+2 = {}^-8$, and so $x = {}^-10$
2. (a) $A = 4\frac{1}{2}$
   (b) from $2 = 6 - 12/r$
   $12/r = 4$ (as $2 = 6-4$)
   and so $r = 3$ (as $12/3 = 4$)
3. (a) When $r = 5$, $V = 24$. The maximum safe speed is 24 kph.
   (b) Substituting 36 for V we have
   $36 = \frac{12}{5}\sqrt{(20r)}$
   $\boxed{\text{M} \frac{5}{12}}$ $\sqrt{(20r)} = 15$
   $\boxed{\text{square}}$ $20r = 225$
   $\boxed{\text{M} \frac{1}{20}}$ $r = 11.25$. The required radius is $11\frac{1}{4}$ m.

4. (a) $F = 13\frac{1}{2}$
   (b) From $48 = 4v^2/3$
   $4v^2 = 144$
   $v^2 = 36$
   and so $v = 6$
   (or $^-6$ if it is sensible to have a negative answer).

5. (a) $T = \pm 1\frac{1}{2}$
   (b) From $4 = \dfrac{12}{\sqrt{c}}$
   $\sqrt{c} = 3$ (as $12/3 = 4$)
   and so $c = 9$

2.3 Changing the subject of a formula

1. (a) m (b) A (c) V (d) F (e) T

Exercise C

1. (a) $b = A/3$ (b) $I = V/R$ (c) $f = (v-u)/t$
   (d) $x = 2m - y$ (e) $v = \sqrt{\dfrac{Fr}{m}}$ (f) $r = \dfrac{mv^2}{F}$
   (g) $u = \sqrt{(v^2 - 2as)}$ (h) $v = \dfrac{2s}{t} - u$ or $v = \dfrac{2s - ut}{t}$

2    $L = g\left(\dfrac{T}{2\pi}\right)^2$ or $L = \dfrac{gT^2}{4\pi^2}$. Hence $L = 0.36$.

3    $h = \dfrac{S - 2\pi r^2}{2\pi r}$; and so $h = 4.68$ (approximately)

4   (a) $A = \pi(R^2 - r^2)$ (b) $R = \sqrt{\left(\dfrac{A}{\pi} + r^2\right)}$ (c) $A = 29.9$ (d) $R = 1.13$

5    $E = mc^2$, and therefore $c = \sqrt{\dfrac{E}{m}}$; hence $c = 3 \times 10^{10}$

2.4 Further examples

1   x is the subject in (b): $x = 3y - 7$.
In (a) the rearranged formula will be $x = \dfrac{by}{c-a}$,
and in (c) the rearranged formula is $x = \tfrac{1}{2}y^2$.

2    $Q = Ps - Pt$
$Q = P(s - t)$
$P = \dfrac{Q}{s - t}$

Exercise D

1    $p = \dfrac{a}{3+q}$ 5 $P = \dfrac{100A}{(100 + RT)}$

2    $a = \dfrac{s}{2c + 4b}$ or $a = \dfrac{s}{2(c + 2b)}$ 6 $x = \dfrac{2z}{1 - z}$

3    $x = \dfrac{5}{y - 1}$ 7 $f = \dfrac{uv}{u + v}$

4    $m = \dfrac{2p}{v^2 - u^2}$ 8 $t = \dfrac{3 - 2w}{2 - 3w}$ or $t = \dfrac{2w - 3}{3w - 2}$

Post-test

1    $a = 180 - \dfrac{360}{n}$; giving (a) $a = 156$ and (b) $n = 9$.
The last answer is obtained from the reversed flow diagram
$a \longrightarrow \boxed{\text{S from 180}} \longrightarrow \boxed{\text{D into 360}} \longrightarrow n$

2    $x = \tfrac{1}{2}(180 - v)$ giving $x = 72$

3   (a) $S = 87.9$ or 88.0 (b) $L = 4.98$ or 4.97

4    $n = v \div \sqrt{(a^2 + x^2)}$ gives $n = 5.17$

5   The excess amount is $(W - 20)$ kg.
The cost will be $\$7 \times (W - 20)$.
Hence $C = 7(W - 20)$. Answer (b) is correct.

3 Gradients

Objectives

This is what you should be able to do after studying this chapter.
(1) Calculate the gradient of a straight-line graph.
(2) Distinguish between positive and negative gradients.
(3) Understand what is meant by 'the rate of change of one quantity with respect to another' and work it out from the gradient of the graph relating the two quantities.
(4) Find the gradient of a linear algebraic function.
(5) Estimate the gradient at a point on a curve.
(6) Relate the distance, speed and acceleration on distance–time and speed–time graphs.

Pre-test

> **1** Express $12\frac{1}{2}\%$ as a fraction, and $\frac{3}{8}$ as a percentage.

2 Draw a graph for the following values of x and y.

x	0.5	1.0	1.5	2.0	2.5
y	1.9	2.5	3.1	3.7	4.3

Where does the graph cut the y-axis?

3 Which of the following functions will give a straight-line graph?
(a) $2y = 3x - 1$ (b) $f: x \to 2x^2 - 1$ (c) $x \to x + \frac{2}{x}$
(d) $y = 3 \cos x°$ (e) $4x + 3y = 7$
Sketch the graphs of those that do.

4 O is $(0, 0)$, A is $(4, 0)$, B is $(6, 0)$, P is $(4, 2)$ and Q is $(6, 3)$.
(a) Draw a diagram to show triangles OAP and OBQ.
(b) What is the connection between the two triangles?
(c) Write down a relation between the sides of the two triangles.

5   Complete the table below. (The results will be used in question **3** of the Assignment)

t	0	1	2	3	4	5	6	7	8	9	10
30t	0	30	60						240		
sin (30t)°	0	0.5	0.87						−0.87		
4 sin (30t)°	0	2.0	3.5						−3.5		

3.1 Rates of change

When a brass rod is heated it expands, and the results of an experiment are shown in the table below:

Temperature (°C)	20	40	60	80	100	125	150
Length (cm)	10.02	10.04	10.06	10.08	10.10	10.125	10.150

Figure 1 shows some of these results pictorially.

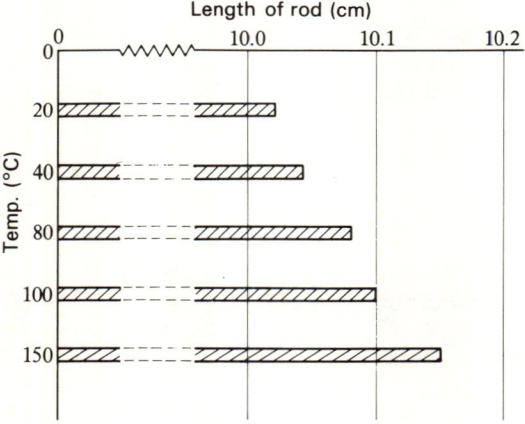

Figure 1

1   If the rod is heated from a temperature of 40 °C to a temperature of 100 °C, work out the following.
(a) The increase in length (in cm).
(b) The increase in temperature (in °C).
(c) $\dfrac{\text{The increase in length}}{\text{The increase in temperature}}$.

From this last result we say that, overall, the length is increasing by 0.001 cm for every degree C that the temperature increases, or that the length is increasing at *the rate of* 0.001 cm per degree C (written as 0.001 cm/°C).

Another way of putting this is to say that 'The average rate of change of length with respect to temperature is 0.001 cm/°C.'

> **2** The result above was calculated for a change of temperature from 40 °C to 100 °C. From the table, work out the average rate of change of length with respect to temperature for the following temperature changes.
(a) 20 °C – 40 °C (b) 40 °C – 60 °C (c) 100 °C–125 °C (d) 125 °C–150 °C
What do you notice?

If we plot the figures given in the table, we obtain the graph shown in Figure 2.

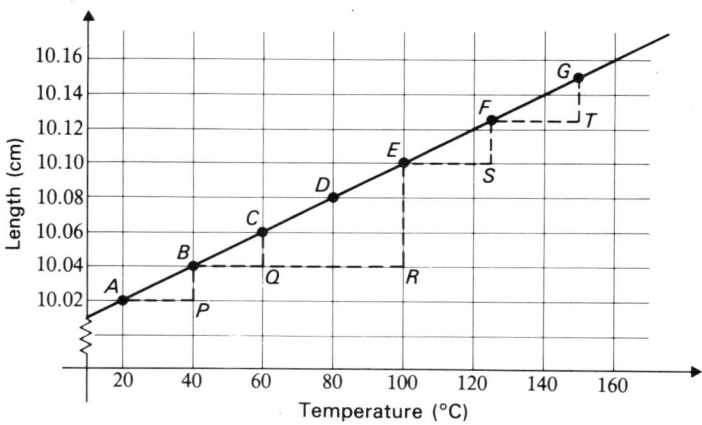

Figure 2

We find that it is a straight-line graph, and we can use it to illustrate the results we obtained earlier. The change in temperature from 40 °C to 100 °C is represented by the line *BR*, and the corresponding change in length by *RE*. Hence the rate of change of the length is represented by the ratio *RE/BR*.

> **3** (a) Which ratios represent the rates of change calculated in **2** above?
(b) What geometrical fact confirms the deduction that you made at the end of **2**?
(c) Would we have obtained different rates of change for different temperature intervals (e.g. 25 °C–35 °C, 50 °C–110 °C)?

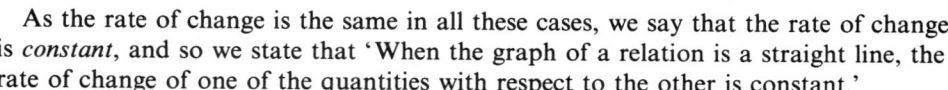

As the rate of change is the same in all these cases, we say that the rate of change is *constant*, and so we state that 'When the graph of a relation is a straight line, the rate of change of one of the quantities with respect to the other is constant.'

'Order' of the variables

In this experiment the 'base' (or independent) variable is temperature. It is the *top* line of the table, and is decided *before* the experiment is performed. The lengths *depend* on the temperatures chosen, and are not known until the experiment has been carried out.

When showing the results graphically, the first (base, or independent) variable goes *across* the page, and the other one up the page.

When stating a rate of change it is the rate of change of the *dependent* variable with respect to the independent variable, 'with respect to' implying that we are taking the independent variable (temperature in this example) as our base or 'yardstick'.

▷ 4 In a survey, the temperatures at given depths below the surface of a lake were recorded with the following results.
Depth below surface (m)   ...
Temperature (°C)   ...
(a) How would you label the axes for a graph?
(b) What rate of change would you calculate, and in what units?

Exercise A

▷ 1 The results of the experiment quoted above were as shown in the following table.

Depth (m)	15	30	45	60	90	150
Temperature (°C)	1.0	1.3	1.6	1.9	2.5	3.7

(a) Represent this information graphically.
(b) Calculate the appropriate rate of change.
(c) Estimate the temperature at the surface.

▷ 2 The temperature of oil being heated is measured at various times with the following results.

Time (s)	0	10	20	30	40	50	60
Temperature (°C)	10	12.1	14.3	16.0	17.7	19.2	20.0

(a) Show these results graphically.
(b) Calculate the average rate of change for the following intervals of time 0 – 10 s, 20 – 40 s, 50 – 60 s.
(c) When is the rate of change greatest?

3.2 Gradients

Surveyor's gradients

The road signs in Figure 3 tell us that we are approaching a steep hill. The one in Figure 3(a) tells us that we are about to go *up* a hill of gradient (or slope) 1 in 4, by which we usually assume that the road climbs 1 metre (vertically) for every 4 metres travelled along the road. The one in Figure 3(b) tells us that we are about to go *down* a hill of gradient 20% (i.e. 20 per 100); that is the road drops 20 m (vertically) for every 100 m along the road.

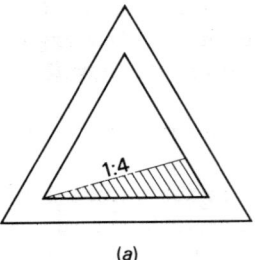 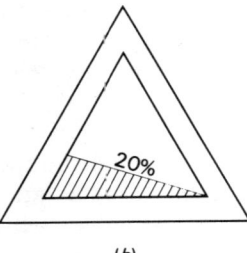

(a)　　　　　　　　(b)

Figure 3

> 1 (a) What is '20%' in the form '1 in ...'?
   (b) Which is steeper, 1 in 5 or 1 in 8? Express both of these as percentages.
   (c) Which is steeper, 10% or 15%? Express these in the form '1 in n' writing n correct to the nearest whole number.
   (d) Which method do you think is the more useful?
   (e) How do the signs distinguish between 'uphill' and 'downhill'?
   (f) If the gradient is 1 in 4 all the way up a hill, does it become steeper, less steep, or remain at the same 'steepness'?

A gradient of 1 in 4 is the same as a gradient of 25%. But 25% is the same as $\frac{1}{4}$, and so an alternative way of writing 'a gradient of 1 in 4' is to write 'a gradient of $\frac{1}{4}$'.

If we draw a *scale* cross-section of such a road (i.e. one for which the same scale is used horizontally and vertically), as in Figure 4, we see that $\frac{1}{4}$ is the sine of the angle with the horizontal; i.e. $\sin \theta° = $ gradient.

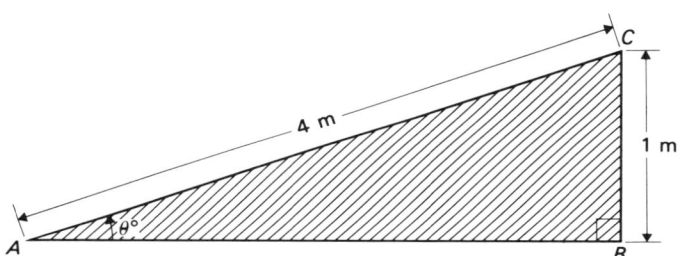

Figure 4

In these cases we could say that the gradient measures the rate of change of the (vertical) rise in the road with respect to the distance measured *along* the road.

Such gradients are usually called surveyor's or geographical gradients.

Mathematical gradients

However, if we regarded Figure 4 as a graph to show the relation between vertical height and *horizontal* distance, we should measure the rate of change by the ratio BC/AB, regarding it as the rate of change of (vertical) height with respect to the distance measured *horizontally*.

The ratio BC/AB measures the *tangent* of the angle, and we use the phrase *mathematical gradient* in this case. That is, mathematical gradient = $\tan \theta°$.

> 1 (a) In practice, few roads are steeper than '1 in 4'.
      If this is a surveyor's gradient, what is the value of θ?
      If this is a mathematical gradient, what is θ?
   (b) What is the difference between the two possible values of θ when the gradient is 10%?
   (c) A mountain path is inclined at an angle of 20° to the horizontal. Write down the surveyor's gradient, and the mathematical gradient of the path, in each case giving it both as a percentage (to the nearest whole number) and in the form '1 in n', where n is correct to 1 decimal place.

45

Gradients on coordinate graphs

Unless otherwise stated, we shall assume from now onwards that we are talking about mathematical gradients when we use the word gradient.

▷ 1 What is the gradient of each of the lines in Figure 5?

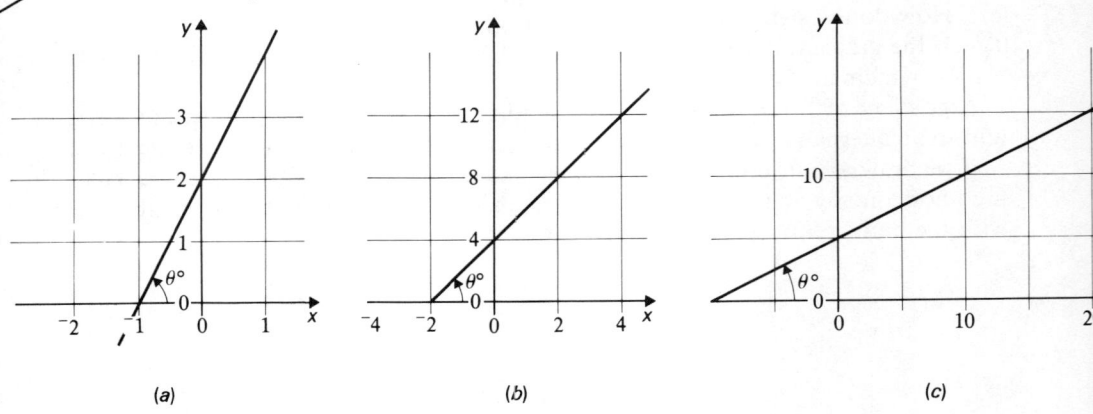

Figure 5

Is the gradient the same at all points on any one line?

▷ 2 Now use your protractor to measure the angle $\theta°$ in each case, and write down the value of tan $\theta°$. What do you notice?

If there is any doubt, remember that the gradient is in terms of the figures marked on the axes, *not* the lengths that are used to represent them. These can be different if the scales are changed. Thus, in Figure 5 the gradients are (a) 2 (b) 2 (c) ½.

If $P(a, b)$ and $Q(c, d)$ are *any* two points on a straight line, then the gradient of the line is given by

$$\text{gradient} = m = \frac{\text{increase in the } y\text{-value}}{\text{increase in the } x\text{-value}} = \frac{d-b}{c-a} \quad \text{(see Figure 6)}.$$

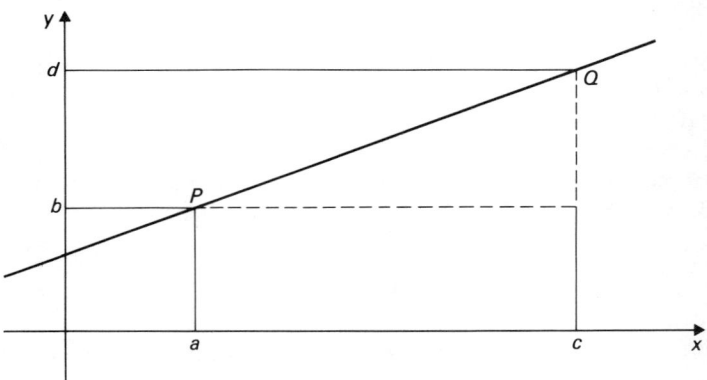

Figure 6

(The letter m is often used to stand for the gradient of a line.)

Negative gradients

What is the gradient of the line in Figure 7? If we think of the line as going from A

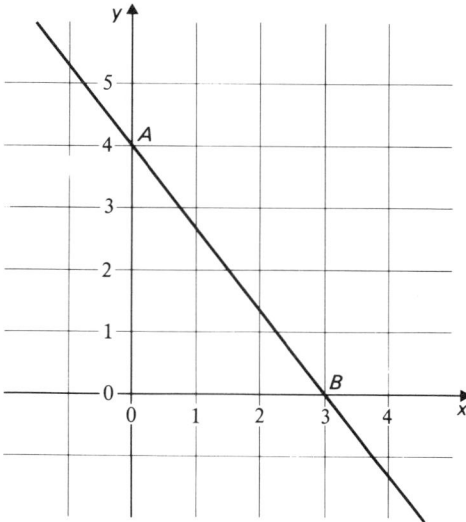

Figure 7

to B, and apply the formula of the previous section, we have

$$m = \frac{\text{change in } y\text{-value}}{\text{change in } x\text{-value}} = \frac{0-4}{3-0} = \frac{-4}{3}.$$

> 1   Does it make any difference if we think of the line as going from B to A?

2   It is probably simplest to think of *all* lines as going across the page from left to right: if, when travelling this way, the line goes up(hill) then its gradient is positive; if it goes down(hill) then its gradient is negative. Does this agree with the convention used on 'steep hill' signs?

3   What is the gradient of the x-axis? and of the y-axis?

4   Is the formula 'gradient $= \tan \theta$' still true for lines with a negative gradient?

Gradients and rates of change

A rate of change is a physical quantity, and must be measured in appropriate units. Its numerical value may change with change of units: thus a speed of 1 m/s is the same speed as one of 3.6 kph. Both express the same *rate* of change of distance with respect to time.

    A gradient is just a number. If a relationship is graphed, the gradient of the graph is the numerical value of the rate of change (when the rate of change is measured in the units used on the graph).

Exercise B

1. (a) Draw x- and y-axes from ⁻3 to ⁺9 and plot the following points:
      A(0, 2) B(6, 4) C(1, 9) D(9, ⁻1) E(⁻3, 1) F(⁻1, ⁻3) G(0, ⁻1)
   (b) Write down the gradients of the following lines:
      AB, AC, AD, AE, AF, AG, BA, BC, BD, EA, EF, FG, GD

2. Draw x- and y-axes from ⁻3 to ⁺3, and through the origin draw lines with the following gradients:

$$3, \tfrac{1}{2}, -2, 0, -\tfrac{1}{4}, -1$$

3. The temperature of a liquid was noted at various times with the following results.

Time (min)	0	1	2	3	4	5	6
Temperature (°C)	12	17	22	27	30	33	36

What can you say about the changes in the temperature?

3.3 Gradient of a linear function

A linear function is defined as one whose graph is a straight line. Therefore its rate of change (if it is applied to a physical situation), or its gradient, is constant.

1. Complete the tables below, and so graph the functions.
   (a) $f: x \to 2x+3$
       (i.e. $y = 2x+3$)
   (b) $g: x \to 1-\tfrac{1}{2}x$
       (i.e. $y = 1-\tfrac{1}{2}x$)
   (c) $h: x \to \tfrac{3}{4}x-\tfrac{1}{4}$
       (i.e. $y = \tfrac{3}{4}x-\tfrac{1}{4}$)

x	0	2	4
y			

x	⁻4	0	⁺4
y			

x	0	3	6
y			

What is the gradient of the line in each case?

From these results you should be able to see that for a linear function in the form

$$f: x \to mx+c$$

the coefficient of x (i.e. the value of m) is the gradient of the graph of the function.

2. On one graph (for values of x from ⁻3 to ⁺3) draw the graphs of $2y = x-3$, $2y = x$ and $2y = x+1$.
   What is the gradient of each of these lines? What do you notice about the lines? What must be done to each of these equations so that the coefficient of x is the gradient of the line?

3. What meaning can be given to the value of c when the equation of a straight line is put in the form $y = mx+c$?
   Write down the equations of the following lines.
   (a) The one that has a gradient of 3 and goes through the point (0, 5).
   (b) The one that is parallel to $y = 3-2x$, and goes through the point (4, 0).

Exercise C

1. Write down the gradients of the following lines (do *not* draw their graphs).
   (a) $y = 5x - 6$
   (b) $y = 1 - \frac{3}{4}x$
   (c) $y = 4 + \frac{1}{2}x$
   (d) $3y = x - 7$
   (e) $\frac{1}{2}y = x + 6$
   (f) $\frac{1}{2}y = \frac{3}{4}x - \frac{1}{4}$
   (g) $x + y = 8$
   (h) $x + 2y = 10$
   (i) $x - 5y = 1$

2. Write down the equations of the lines with the following gradients, through the given points.
   (a) Gradient 4, through the point (0, 3)
   (b) $m = ^-3$ (0, 1)
   (c) $m = \frac{1}{2}$ (0, $^-$20)
   (d) $m = 1\frac{1}{2}$ (1, 1)
   (e) $m = ^-10$ (5, 0)
   (f) $m = 1.44$ (1, 0)

3. To verify Hooke's Law, a number of different masses are hung on the end of a light length of wire hanging vertically, and the length noted in each case.

Mass (kg)	5	10	20	25
Length (cm)	51.5	53.0	56.0	57.5

   (a) Show these results on a graph.
   (b) What is the gradient of the graph? Is it constant?
   (c) What rate of change is measured by the gradient?
   (d) Estimate the unstretched length (d cm) of the wire.
   (e) Write down formulas for the stretch of the wire (D cm) in terms of d and the total length (L cm), and in terms of the mass (M kg).
   (f) Write down a formula for L in terms of M.

3.4 Gradient of a non-linear function

Figure 8 shows the relation between the diameters and areas of circles of various radii.

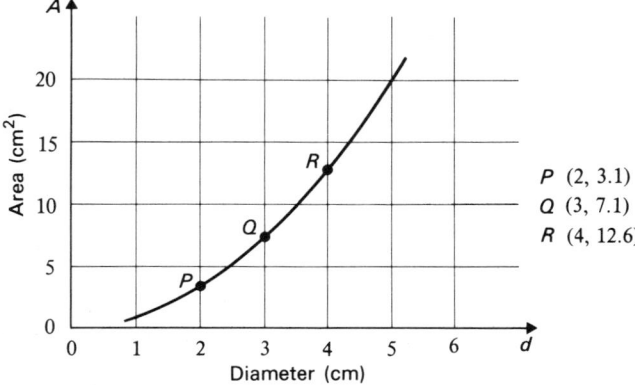

Figure 8

▷ 1 Is the gradient of this graph constant? If not, is it increasing or decreasing? Will the gradient ever be zero or infinite (for $d > 0$)?

In such examples as the one above we cannot speak about *the* gradient, or *the* rate of change, as they are changing continuously. We *can* do two things.

First, we can consider the average gradient for a given interval.

2 What is the average rate of increase in area, with respect to the diameter, as d goes from 2 to 4?

Secondly, we can estimate the gradient at one particular point. But what, for example, do we mean by the gradient at the point Q (whose coordinates are $(3, 7.1)$) when the gradient is continuously increasing. If, by some means, we could halt this *increase* as soon as d reached 3 (so that from Q onwards the gradient was constant, with the effect that the graph would now go from Q to T instead of from Q to R) then the gradient of the straight line QT shows us what the gradient of the curve is *at* Q (see Figure 9(a)). This can be illustrated by thinking of the graph as a model railway track, of which PQ and QR are two curved sections, and replacing the curve QR by a 'straight' section QT. (See Figure 9(b)).

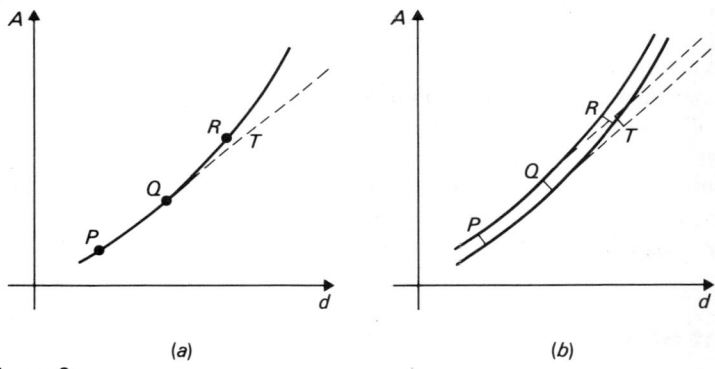

(a)   (b)

Figure 9

Geometrically, QT is the tangent to the curve at the point Q. In practice this can be obtained in the following two ways.

▷ 3 *By drawing*: Make an accurate copy of Figure 9(a). Two more points that will help with the plotting are $U(2.5, 4.9)$ and $V(3.5, 9.6)$. Draw the tangent at Q (some people find it easier to draw the 'normal' first – that is, the line at right angles to the curve at Q – and then draw the tangent at right angles to that), and calculate its gradient. (If you let T be the point on the tangent whose x-coordinate is 4 or 5, the calculation is not very difficult!)

▷ 4 *By calculation*: Calculate the gradients of the chords PQ, QR and PR. Calculate the gradients of the chords UQ, QV and UV. If you have a calculator, calculate more accurate values of A for the points $F(2.9, \ldots)$ $Q(3, \ldots)$ and $N(3.1, \ldots)$ from the formula $A = \frac{1}{4}\pi d^2$, and hence calculate the gradients of the chords FQ, QN and FN.

From all these results you should agree that the gradient of the tangent at Q is about 4.7.

Exercise D

1. (a) Complete the table below for the function $y = x^3$.

x	0	1	1.9	2	2.1	2.5
y	0	1	6.86			

   (b) Represent these values on a graph.
   (c) Write down the average gradient for the intervals $x = 1$ to $x = 2$; $x = 2$ to $x = 2.5$.
   (d) Estimate the gradient when $x = 2$ by drawing the tangent, and by calculation, using the values evaluated for the table above.

2. The table below gives information about a pig being fattened for market.

Day	0	10	20	30	40	50	60	70	80	90
Mass (kg)	2	6	20	28	39	49	56	70	84	95

   (a) Represent this information on a graph.
   (b) What is the rate of increase on the 40th day?
   (c) What is the average rate of increase over the following periods: day 0 to day 90; day 0 to day 20; day 20 to day 60; day 60 to day 90?
   During which of these periods was the mass increasing the fastest?

3.5 Distance, speed and acceleration

Distance–time graphs (travel graphs)

1. Figure 10 shows the graphical representation of the journey of a cyclist who rides for 20 minutes and then stops. How far does the cyclist travel between (a) $t = 0$ min and $t = 5$ min, (b) $t = 5$ min and $t = 10$ min, and (c) $t = 10$ min and $t = 20$ min?

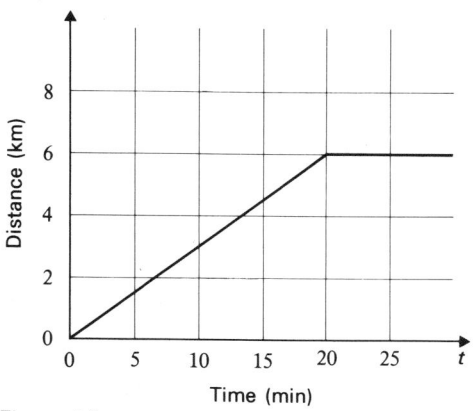

Figure 10

   (d) What are his average speeds during these intervals (in km per minute)?
   (e) What is the gradient of the graph between $t = 0$ min and $t = 20$ min, and for $t \geqslant 20$?

2 Figure 11 shows the progress of another cyclist during an interval of 1 minute.

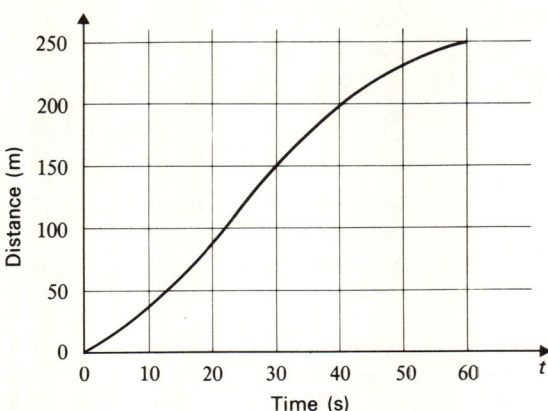

Figure 11

(a) What is the average speed over this period of time?
(b) Draw a line on the graph to show his greatest speed. What is his greatest speed? At what time was he travelling at this speed?

Both these examples show that on a distance–time graph the gradient represents the numerical value of the speed.

Exercise E

 1 Draw an accuracte distance–time graph for the following journey.
(a) A steady speed of 15 m/s for 3 minutes; followed by
(b) a stop for $\frac{1}{2}$ minute, followed by
(c) a steady speed of 10 m/s for 4 minutes.

2 A sledge sliding down a slope travels a distance of s metres in t seconds, where the relation between s and t is
$$s = 10t + t^2.$$

(a) Graph this relation for values of t from 0 to 4.
(b) Calculate the average speed between $t = 0$ and $t = 4$.
(c) Estimate the speed of the sledge when $t = 1$, 2.4 and 3.

Speed–time graphs

 1 Figure 12 shows a five minute train journey between two stations. What happens to the speed (a) for the first 150 s, (b) for the next 100 s, (c) for the remainder of the journey?
What is the gradient of the graph for $0 \leqslant t \leqslant 150$? What rate of change does this show?

This rate of change of speed with respect to time is called acceleration. In scientific work, the unit of time in the speed part and the time part should be the same (and would normally be seconds, i.e. an acceleration would be quoted in cm/s

52

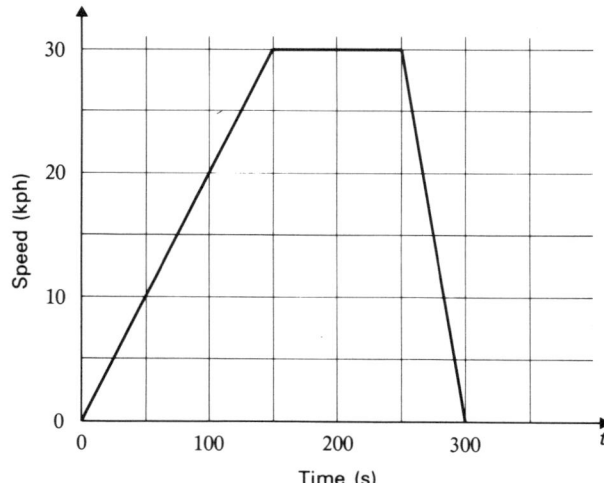

Figure 12

per second, cm/s² or cm s⁻², or m/s²). In practical work it is permissible to mix the units. A realistic unit for a car's acceleration would be kph per second, for example.

> **2** Referring to Figure 12 again, what is the gradient and acceleration for the intervals (a) $150 \leqslant t \leqslant 250$ (b) $250 \leqslant t \leqslant 300$?

> **3** Figure 13 is a more realistic graph of an actual journey.

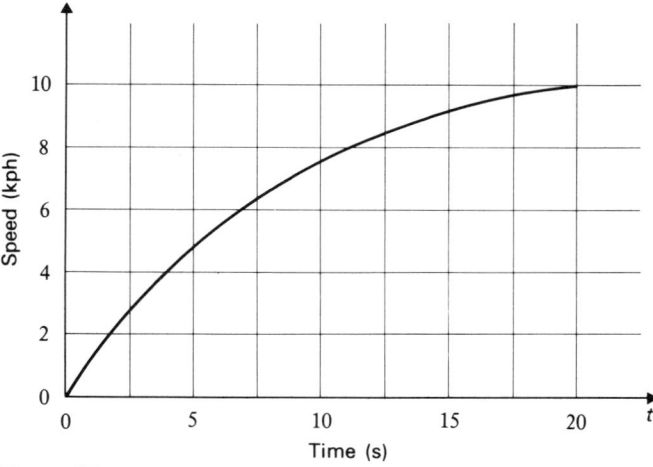

Figure 13

(a) What does the gradient of the graph represent?
(b) What is the average gradient over the 20 s interval?
(c) What is the greatest acceleration? When does it occur?

It should now be clear that the gradient of a speed-time curve represents the acceleration.

53

Exercise F

1. Figure 14 shows one acceleration–time graph, and three speed–time graphs.

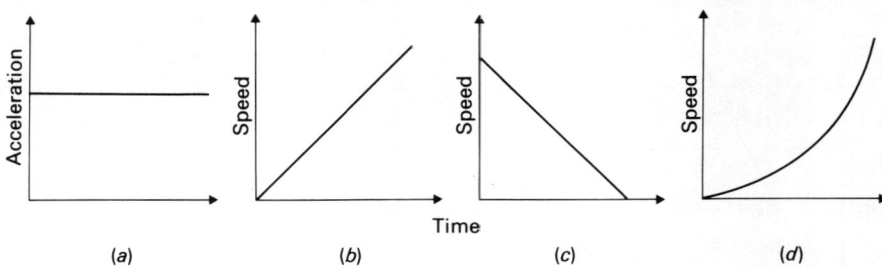

Figure 14

(a) Which speed–time graph corresponds to the acceleration–time graph?
(b) Sketch acceleration–time graphs to correspond to the other two speed–time graphs.

2. What is the acceleration of a train that does the following
   (a) Increases its speed steadily from 0 m/s to 20 m/s in 200 s
   (b) Increases its speed steadily from 15 m/s to 24 m/s in 2 min
   (c) Slows to rest steadily from 40 m/s in 160 s

3. Draw an accurate acceleration–time graph for the train journey represented by Figure 12.

Summary

(1) The gradient of a straight line is the ratio of the vertical displacement to the corresponding horizontal displacement.

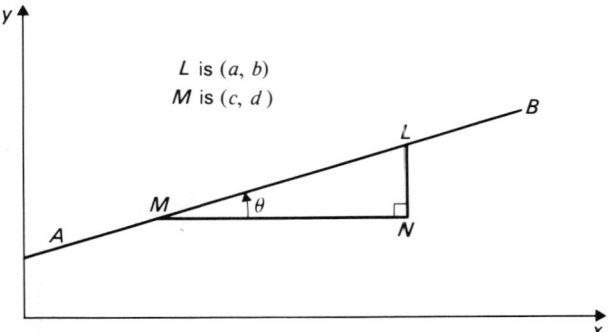

Figure 15

In Figure 15 the gradient of the line AB is

$$m = \frac{LN}{MN} = \frac{b-d}{a-c} = \tan \theta$$

54

(provided both axes are drawn to the same scale). Note that the 'lengths' *LN* and *MN* are in terms of the coordinates of *L* and *M*, or the quantities represented by *LN* and *MN*, *not* the actual lengths on the page of a book.

(2) When the line slopes downwards (going from left to right) the gradient is negative. The gradient of the *x*-axis (and lines parallel to it) is 0. The gradient of the *y*-axis (and lines parallel to it) is infinite.

(3) For any relation $x \to y$, the gradient of the graph is the rate of change of *y* with respect to *x* measured in the units used on the axes of the graph. The gradient of the graph in Figure 16 is $40/10 = 4$. It represents a rate of change of volume with respect to time of 4 cm³/min.

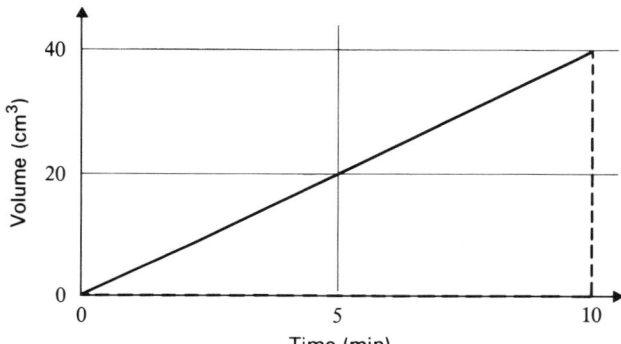

Figure 16

(4) For a linear function of the form $f: x \to mx+c$ (or an equation of the form $y = mx+c$) the gradient of the graph is equal to *m*, the coefficient of *x*.

(5) For a non-linear function (i.e. one whose graph is curved) the gradient is continuously changing. We can measure either (a) the average gradient over an interval (of *x*) or (b) the gradient at a point of the curve. The latter is found either by drawing a *tangent* to the curve at the required point, or by calculating the average gradients of chords in the region of the point, and estimating a limiting value.

(6) The gradient of a distance–time graph measures the speed, and the gradient of a speed–time graph measures the acceleration.

Post-test

1 In Figure 17 *A* is the point (5, 4), *B* is (⁻3, 2), *C* is (6, 2), *D* is (6, ⁻1). Calculate the gradients of *AB*, *AC*, *BC*, *CD*, and *DB*.

2 Write down the gradients of the following, and hence *sketch* their graphs.
(a) $f: x \to 2+5x$ (b) $g: x \to 3-8x$ (c) $4y+3x = 24$

3 The boiling point of water at various heights above sea-level is given in the table.

Height (km)	0	1	2	3
Boiling point (°C)	100	96.9	93.8	90.7

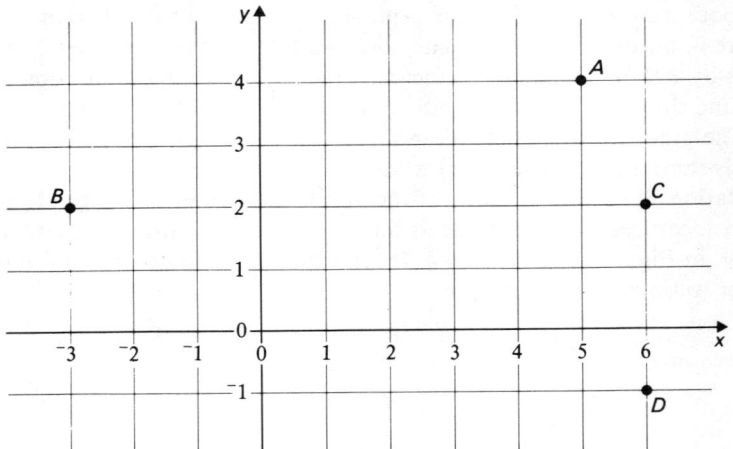

Figure 17

(a) Would these figures give a straight-line graph?
(b) Calculate the rate of change of boiling point with respect to height.
(c) In a well-known experiment, water was found to boil at 85 °C at the top of Mont Blanc. Estimate the height of Mont Blanc based on the data of this question.

4 Figure 18 is a distance–time graph of an object for an interval of 4 s. Estimate the speed after 1, 2, 3 and 4 s.

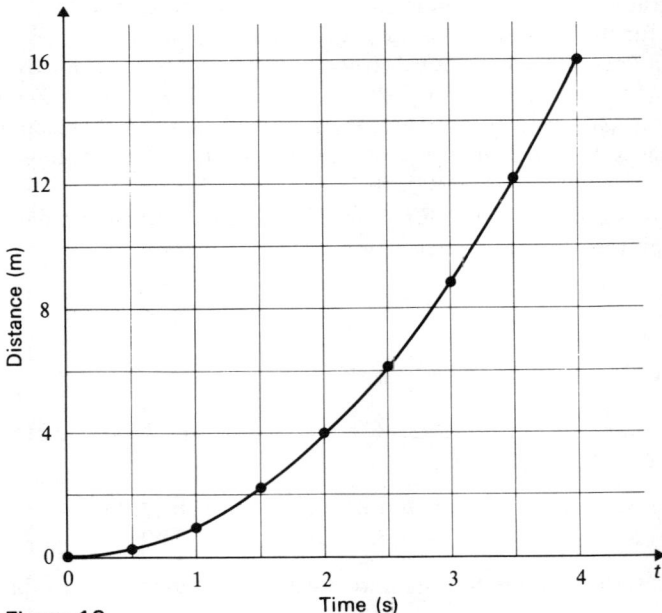

Figure 18

Assignment

1. (a) Draw the graph of $y = 1/x^2$ for values of x from $^-2$ to $^+5$. (Calculate values of y for $x = ^-2, ^-1\frac{1}{2}, ^-1, 1, 1\frac{1}{2}, 2, 2\frac{1}{2}, 3, 4$ and 5.)
   (b) Draw the tangent to the curve at the point $(1, 1)$ and hence estimate the gradient (at that point) of the curve.
   (c) If the equation of the tangent at $(1, 1)$ is $y = mx + c$, state the value of m, and use the fact that the tangent goes through the point $(1, 1)$ to find the value of c.
   (d) What is the gradient of the curve at the point $(^-1, 1)$? What is the equation of the tangent at the point $(^-1, 1)$?

2. A triangle has vertices $F(2, 1)$, $G(5, 2)$, $H(3, 3)$.
   (a) Write down the gradients of the sides of the triangle.
   (b) What is the product of the gradients of FH and GH?
   (c) What is the relation between the sides FH and GH? (Plot the points on squared paper.)
   (d) Find the equation to the side FG.

3. At noon on a certain day, the water in a harbour was at MSL (mean sea level, i.e. half-way between high and low tide). At a time t hours later, the height of the water about this level (h metres) is given by the formula

   $$h = 4 \sin (30t)°.$$

   Calculate (a) the height above MSL at 2.20 p.m. and 3.00 p.m., (b) the average rate at which the tide rose between 2.20 p.m. and 3.00 p.m. Also, estimate (from a graph drawn for values of t from 0 to 10), (c) the time of high tide, (d) the time (between midday and 10.00 p.m.) when the tide is falling the quickest, (e) the rate at which the tide is then falling.
   (Use the results to question **5** of the pre-test to draw the graph.)

Answers

Pre-test

1. $12\frac{1}{2}\% = \frac{1}{8}$, $\frac{3}{8} = 37\frac{1}{2}\%$.
2. The graph cuts the y-axis at $(0, 1.3)$ (see Figure A over the page).
3. (a) and (e) will give straight-line graphs (see Figure B over the page).
4. (a) See Figure C over the page.
   (b) The two triangles are similar.
   (c) $\dfrac{OA}{OB} = \dfrac{AP}{BQ} = \dfrac{OP}{OQ} \left(= \dfrac{2}{3}\right)$
   or $OA = \frac{2}{3}OB$, $AP = \frac{2}{3}BQ$, $OP = \frac{2}{3}OQ$.

5.

t	3	4	5	6	7	8	9	10
$30t$	90	120	150	180	210	240	270	300
$\sin (30t)°$	1	0.87	0.50	0	$^-0.50$	$^-0.87$	$^-1$	$^-0.87$
$4 \sin (30t)°$	4	3.5	2.0	0	$^-2.0$	$^-3.5$	$^-4$	$^-3.5$

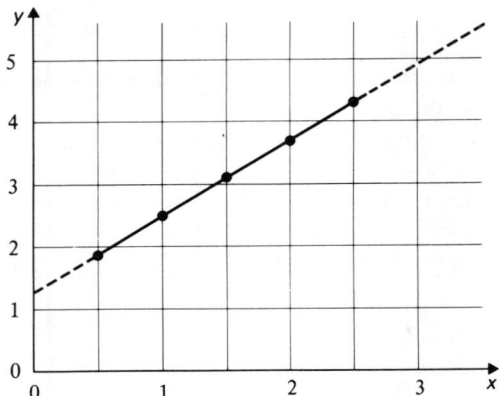
Figure A

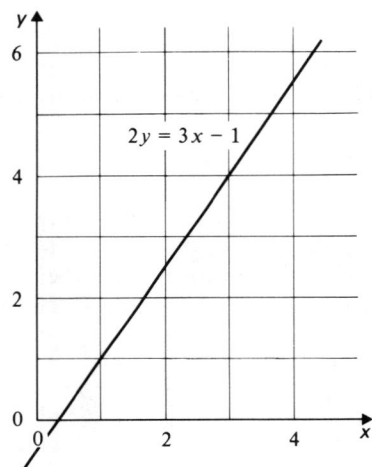
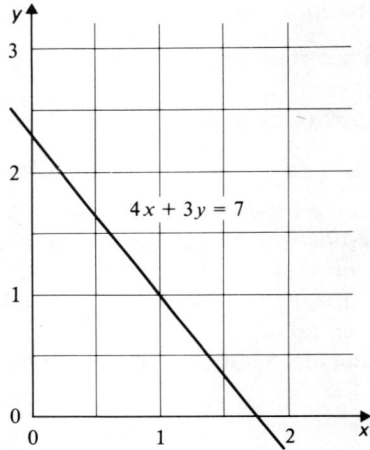
Figure B

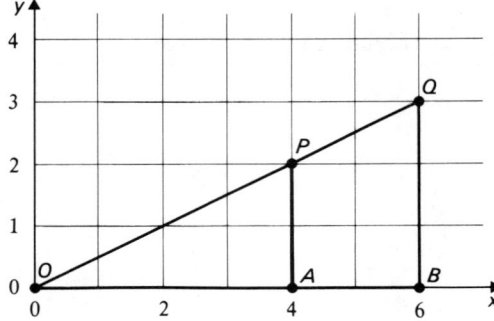
Figure C

3.1 Rates of change

> **1** (a) The increase in length is 0.06 cm.
>   (b) The increase in temperature is 60 °C.
>   (c) $\dfrac{\text{Increase in length}}{\text{Increase in temperature}} = \dfrac{0.06}{60} = 0.001 \text{ cm/°C}$.

> **2** (a) $\dfrac{(10.04 - 10.02) \text{ cm}}{(40 - 20) \text{ °C}} = \dfrac{0.02}{20} = 0.001 \text{ cm/°C}$.
>   (b), (c) and (d) all give 0.001 cm/°C as well.

> **3** (a) 20 °C – 40 °C is shown by $\dfrac{BP}{AP}, \dfrac{CQ}{BQ}, \dfrac{FS}{ES}, \dfrac{GT}{FT}$
>   (b) These are all the ratios of corresponding sides in the *similar* triangles ABP, BCQ, EFS and FGT, and are therefore equal.
>   (c) As $ABC\ldots G$ is a straight line, it doesn't matter where we draw the right-angled triangle, i.e. we shall obtain the same *rate* of change whatever the temperature interval.

'Order' of the variables

> **4** (a) See Figure D.

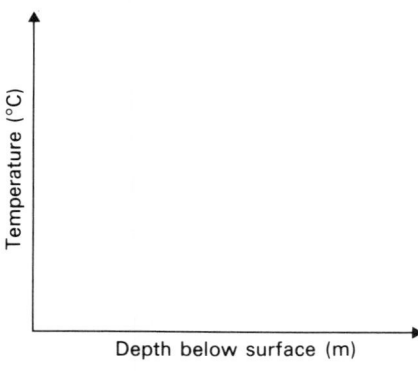

Figure D

>   (b) Rate of change of temperature with respect to depth below surface, measured in °C per metre (or °C/m).

Exercise A

> **1** (a) See Figure E over the page.
>   (b) The rate of change is 0.02 °C/m. (For example, in the dotted triangle, the increase in temperature is 1.5 °C, the increase in depth is 75 m, and so the rate of change of temperature with respect to depth is 1.5 °C/75 m = 0.02 °C/m.)
>   (c) As we *approach* the surface, the temperature is *decreasing* at the rate of 0.02 °C/m.
>   At 15 m the temperature is 1.0 °C; hence at 0 m the temperature is $(1 - 0.02 \times 15) \text{ °C} = 0.7 \text{ °C}$.

> **2** (a) See Figure F over the page.

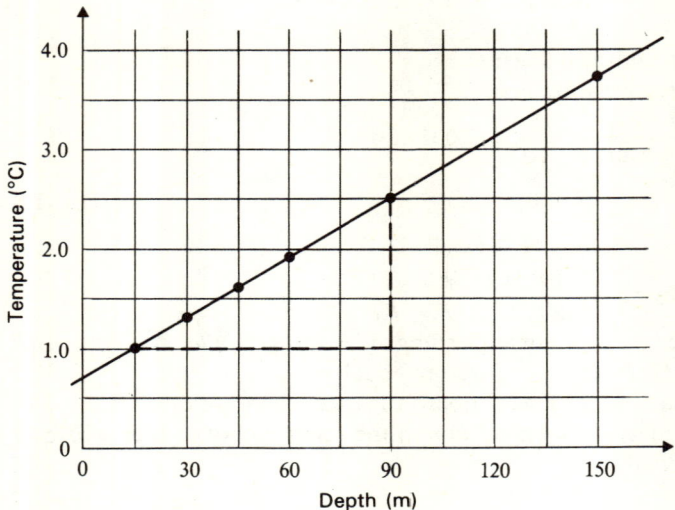

Figure E

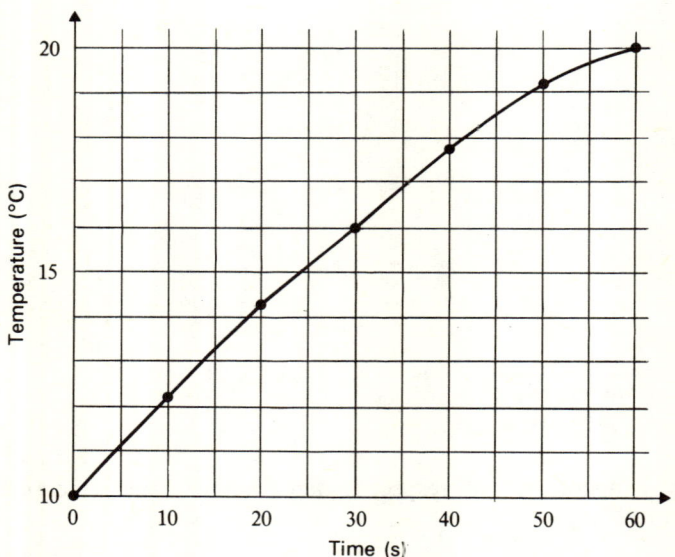

Figure F

(b) For 0–10 s, the average rate of change is 0.21 °C/s, for 20–40 s, the average rate of change is 3.4 ÷ 20 °C/s = 0.17 °C/s, and for 50–60 s, the average rate of change is 0.08 °C/s.

(c) The rate of change is greatest between 10 and 20 seconds; during this time-interval the average rate of change is 0.22 °C/s.

3.2 Gradients

Surveyor's gradients

▶ 1 (a) 1 in 5.
   (b) 1 in 5 is steeper. 1 in 5 is 20%, 1 in 8 is $12\frac{1}{2}$%.
   (c) 15% is steeper. 10% is 1 in 10, 15% is (approximately) 1 in 7.
   (d) The percentage form is probably more useful, as, the *higher* the percentage, the steeper the hill. Also, the use of percentages (even to the nearest whole number) gives a 'wider' range of possible values for the gradients of hills.
   (e) For 'uphill', the slope goes *up* from left to right, for 'downhill' the slope goes down from left to right, i.e. the signs assume that you are travelling from left to right.
   (f) If the gradient is constant, then the 'steepness' is the same at all points on the hill.

Mathematical gradients

▶ 1 (a) If '1 in 4' is a surveyor's gradient, then $\sin\theta° = 0.25$, and so from 3-figure tables $\theta° = 14.5°$.
   If it is a mathematical gradient, then $\tan\theta° = 0.25$ and from 3-figure tables $\theta° = 14.0°$ or $14.1°$.
   In practice, the difference between these angles is not noticeable, and it doesn't really matter whether we regard the gradient of a 'steep' hill as a surveyor's gradient or a mathematical gradient. Also, of course, the expression '1 in 4' is only an approximation, probably covering everything from '1 in $3\frac{3}{4}$' to '1 in $4\frac{1}{4}$'.
   (b) For a gradient of 10%, if $0.1 = \sin\theta°$, then from 3-figure tables $\theta° = 5.7°$ or $5.8°$. If $0.1 = \tan\theta°$, then $\theta° = 5.7°$. The difference is neglible.
   (c) $\sin 20° = 0.342$, and so the surveyor's gradient is 34%, or approximately 1 in 2.9. $\tan 20° = 0.364$, and so the mathematical gradient is 36%, or approximately 1 in 2.7(5).
   $\left(36\% = \dfrac{36}{100} = \dfrac{1}{100/36} = \dfrac{1}{2.75},\right.$
   from reciprocal tables, $\dfrac{1}{36} = 0.0275$; hence $\left.\dfrac{100}{36} = 2.75.\right)$

Gradients on coordinate graphs

▶ 1 (a) 2 (b) 2 (c) $\frac{1}{2}$,
   Yes. All the graphs are *straight lines* and so the gradients are constant.
▶ 2 (a) $\theta° = 63(\frac{1}{2})°$ $(\tan\theta° = 2)$
   (b) $\theta° = 45°$ $(\tan\theta° = 1)$
   (c) $\theta° = 26(\frac{1}{2})°$ $(\tan\theta° = \frac{1}{2})$
   In (a) and (c) the tangent of the angle gives the gradient, but not in (b), because the scale for the y-axis is not the same as that for the x-axis.

Negative gradients

1. No. $m = \dfrac{4-0}{0-3} = -\tfrac{4}{3}$, as before
2. Yes.
3. The gradient of the x-axis is 0, and the gradient of the y-axis is 'infinite'.
4. Yes, because for $90° < \theta° < 180°$ (or $-90° < \theta° < 0°$) $\tan \theta°$ is negative (see Figure G.).

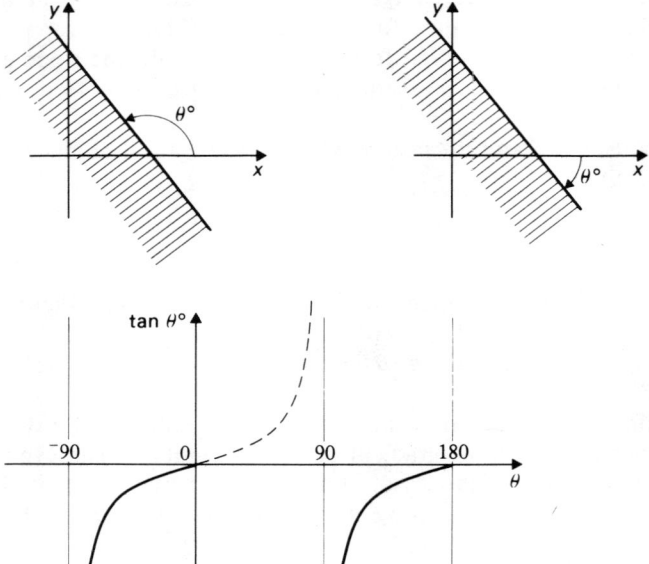

Figure G

Exercise B

1. (a) See Figure H.
   (b) $AB = \tfrac{2}{6} = \tfrac{1}{3}$, $AC = \tfrac{7}{1} = 7$, $AD = (-3)/9 = -\tfrac{1}{3}$, $AE = (-1)/(-3) = \tfrac{1}{3}$, $AF = (-5)/(-1) = 5$, the gradient of AG is 'infinite', $BA = (-2)/(-6) = \tfrac{1}{3}$ (i.e. the same as AB), $BC = -1$, $BD = -\tfrac{5}{3}$, $EA = \tfrac{1}{3}$ (same as AE), $EF = -2$, $FG = 2$, $GD = 0$.
2. See Figure I.
3. During the first three minutes the temperature rose by 5 °C in each interval of 1 min; i.e. the rate of change of temperature with respect to time was 5 °C/min.
   For the next three minutes the rate of change was 3 °C/min. Graphically this is shown in Figure J over the page.

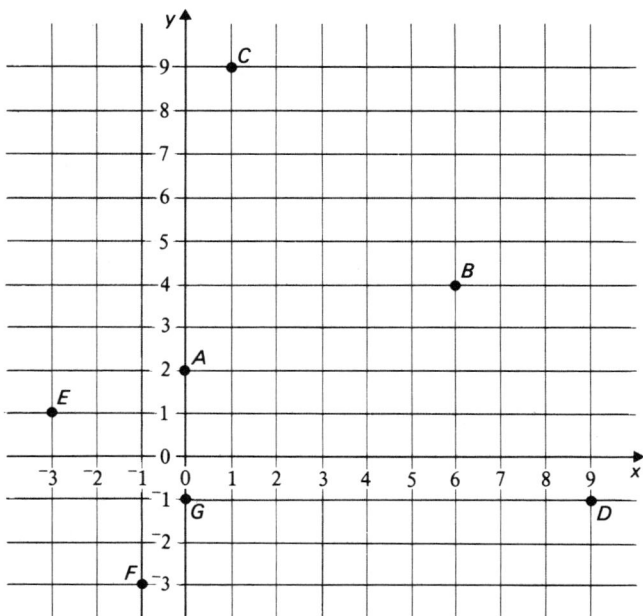

Figure H

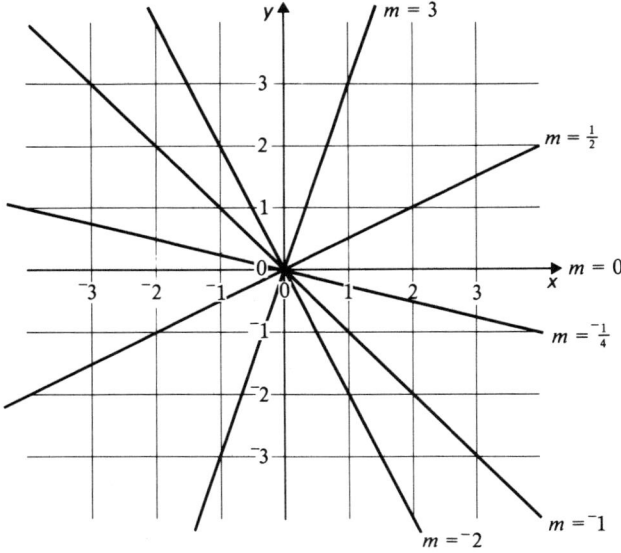

Figure I

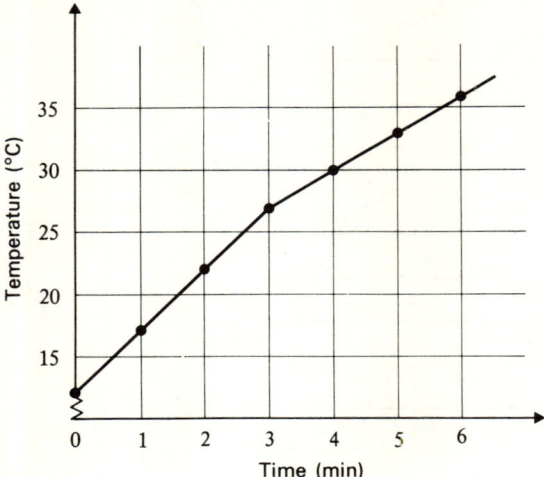

Figure J

3.3 Gradient of a linear function

 1 (a)

x	0	2	4
y	3	7	11

gradient $= \dfrac{8}{4} = 2$

(b)

x	⁻4	0	4
y	3	1	⁻1

gradient $= \dfrac{-4}{8} = -\tfrac{1}{2}$

(c)

x	0	3	6
y	$-\tfrac{1}{4}$	2	$4\tfrac{1}{4}$

gradient $= \dfrac{4\tfrac{1}{2}}{6} = \tfrac{3}{4}$

For plots of these, see Figure K.

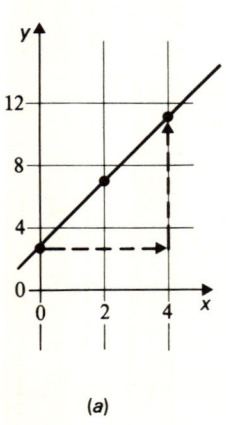

(a)

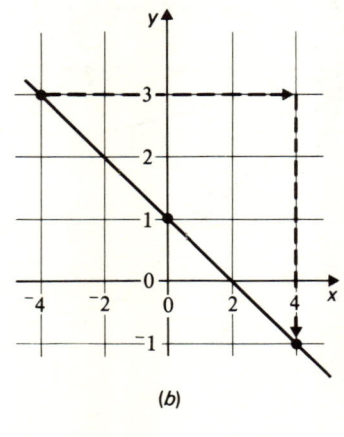

(b)

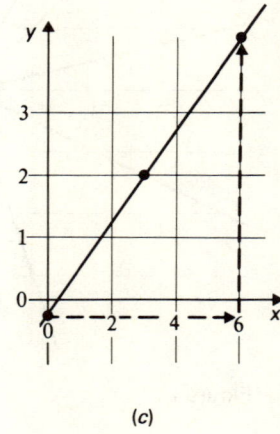

(c)

Figure K

> 2   See Figure L.
Each of these lines has a gradient of $\frac{1}{2}$, and all the lines are parallel.

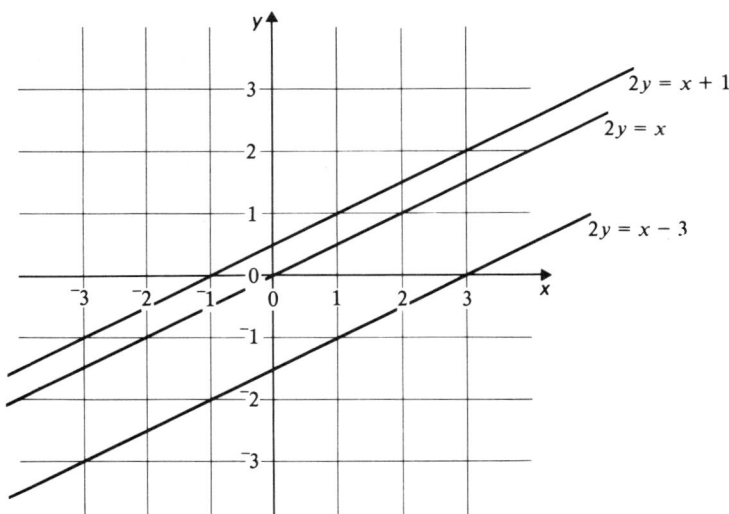

Figure L

If the equations are written in the form $y = \ldots$ (i.e. by dividing both sides of the equation by 2) we have

$$y = \tfrac{1}{2}x - 1\tfrac{1}{2}, \quad y = \tfrac{1}{2}x, \quad \text{and} \quad y = \tfrac{1}{2}x + \tfrac{1}{2}.$$

and in each of these forms the coefficient of x *is* the gradient.

> 3   From the rewritten forms of the equations above, and their graphs we can see that in the equation $y = mx + c$ the 'c' term tells us where the line cuts the y-axis. See, for example, Figure M.

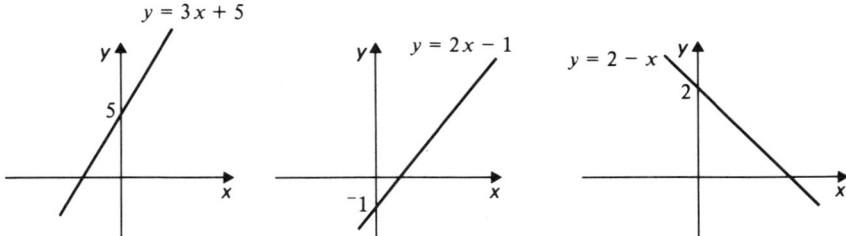

Figure M

(a) $y = 3x + 5$ (see Figure M).
(b) The equation will be of the form $y = k - 2x$ for some value of k. But the line passes through the point $(4, 0)$, (i.e. when $x = 4$ then $y = 0$), and so $k - 2 \times (4) = 0$, or $k = 8$.
Hence the required equation is $y = 8 - 2x$.

Exercise C

1. (a) 5 (b) $-\frac{3}{4}$ (c) $\frac{1}{2}$ (d) $\frac{1}{3}$ (by rewriting the equation as $y = \frac{1}{3}x - 2\frac{1}{3}$) (e) 2 (by rewriting the equation as $y = 2x + 12$) (f) $1\frac{1}{2}$ (g) $^-1$ (rewrite as $y = -x + 8$) (h) $-\frac{1}{2}$ (rewrite as $y = -\frac{1}{2}x + 5$) (i) $\frac{1}{5}$

2. Each equation will be of the form $y = mx + $ something, where m is the gradient of the line.
   (a) $y = 4x + 3$ (b) $y = -3x + 1$ (or $y = 1 - 3x$, or $3x + y = 1$) (c) $y = \frac{1}{2}x - 20$ (which can now be rewritten as $2y = x - 40$) (d) $y = 1\frac{1}{2}x + k$ such that $(1, 1)$ lies on the line. This gives $1 = 1\frac{1}{2} + k$; so k is $-\frac{1}{2}$ and the equation is $y = 1\frac{1}{2}x - \frac{1}{2}$
   (e) $y = k - 10x$ such that when x is 5, y is 0, i.e. $k = 50$, giving $y = 50 - 10x$ (or $10x + y = 50$) (f) $y = 1.44x - 1.44$ (or $y = 1.44(x - 1)$)

3. (a) See Figure N.

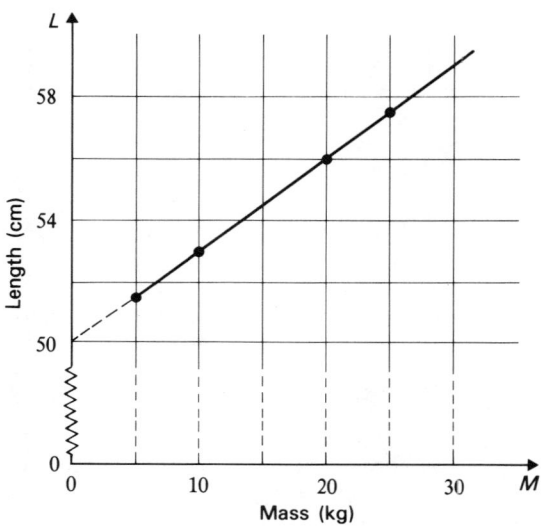

Figure N

(b) The gradient is constant with a value of 0.3.

(c) The gradient measures the rate of change of length with respect to mass. The rate of change is 0.3 cm/kg.

(d) By producing the graph back to the L-axis (as shown by the dotted line) the unstretched length (d cm) appears to be 50 cm. (This assumes that the gradient remains constant as the values of M decrease – in practice this is so in this experiment.)

(e) $D = L - d$ (and hence $L = D + d$); $D = 0.3M$

(f) $L = 0.3M + 50$ (using $L = D + d$)

3.4 Gradient of a non-linear function

1. The gradient of this curve is not constant; it is increasing as x increases. It is zero for $d = 0$, but for all positive, finite values of d the gradient is positive but never infinite.

This may not be easy to accept, as it appears to be increasing 'faster and faster'. But consider a large value, say 1000, for d. The points (1000, 785 398) and (1001, 786 970) are 'adjacent' points on the curve. The average gradient between these two points is 1572, a large, but still finite, gradient!

2 Increase in area $= (12.6 - 3.1)$ cm² $= 9.5$ cm². Average rate of increase in area $= 9.5/2$ cm²/cm $= 4.75$ cm²/cm.

3 See Figure O.

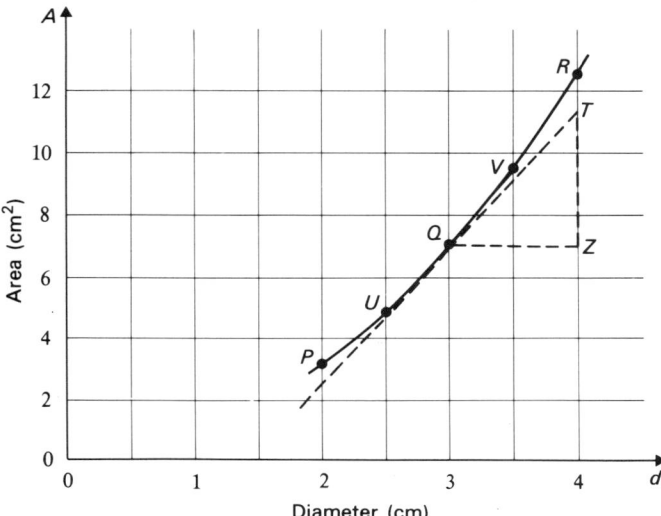

Figure O

$$\text{gradient} = \frac{TZ}{QZ} \approx \frac{4.7}{1} = 4.7$$

4 PQ: gradient is $\dfrac{7.1 - 3.1}{3 - 2} = 4.0$

QR: gradient is 5.5

PR: gradient is $\dfrac{12.6 - 3.1}{4 - 2} = 4.7(5)$.

PQ is less steep than the tangent, QR is steeper, and PR is approximately parallel to the tangent. So our first approximation to the gradient at Q is 4.7(5).

UQ: gradient is $\dfrac{7.1 - 4.9}{3 - 2.5} = 4.4$

QV: gradient is 5.0

UV: gradient is $\dfrac{9.6 - 4.9}{3.5 - 2.5} = 4.7(0)$

Using a calculator F is (2.9, 6.6052...), Q is (3.0, 7.0686...), N is (3.1, 7.5477...)
Gradient of FQ is 4.63..., gradient of QN is 4.79..., and the gradient of FN is 4.71...

Exercise D

1. (a)

x	0	1	1.9	2	2.1	2.5
$y = x^3$	0	1	6.86	8	9.26	15.62

(b) See Figure P.

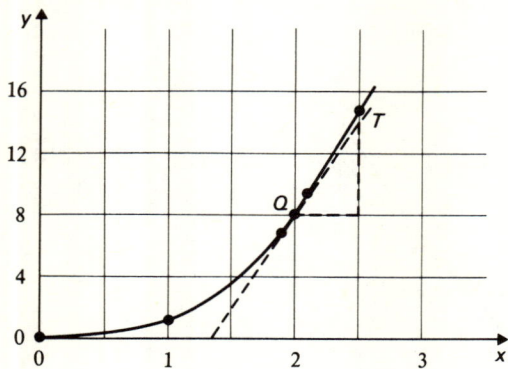

Figure P

(c) For $1 \leq x \leq 2$ the average gradient is 7, and for $2 \leq x \leq 2.5$ the average gradient is
$$\frac{15.62 - 8}{2.5 - 2} = 15.2 \text{ (approximately)}.$$

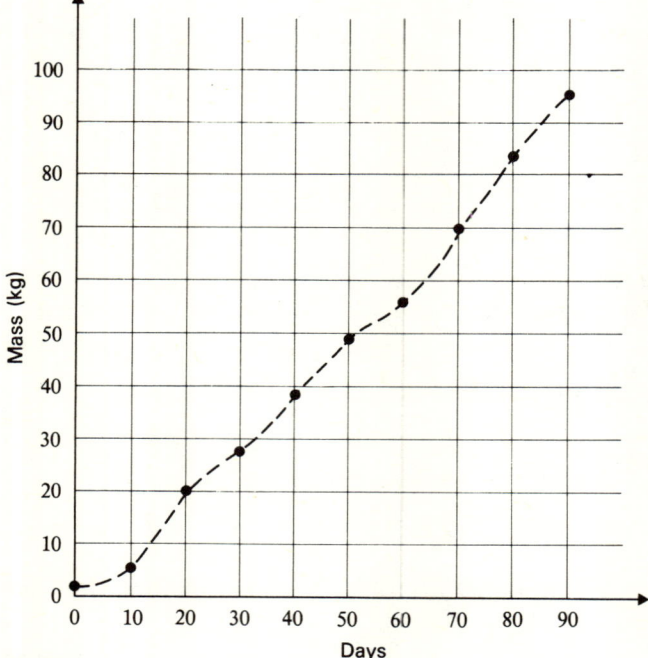

Figure Q

(d) By drawing, T is approximately (2.5, 14), and so the gradient of QT is $\dfrac{14-8}{2.5-2} = 12$. The average gradient for $1.9 \leqslant x \leqslant 2.1$ is $\dfrac{9.26-6.86}{2.1-1.9} = 12$.

These results suggest that the gradient at $x = 2$ is 12.

> 2 (a) See Figure Q on previous page.
(b) Approximately 1 kg per day.
(c) 1.03, 0.9, 0.9, 1.3 (all in kg per day). The mass of the pig was increasing most quickly in the period 'day 60 to day 90'.

3.5 Distance, speed and acceleration

Distance–time graphs

> 1 (a) $1\frac{1}{2}$ km (b) $1\frac{1}{2}$ km (c) 3 km
(d) 0.3 km/min, 0.3 km/min, 0.3 km/min.
(e) For $0 \leqslant t \leqslant 20$ the gradient is 0.3, and for $t > 20$ the gradient is 0.

> 2 Average speed is 250 m/min (or approximately 4 m/s). The greatest speed occurs when $t = 25$ s. By drawing, this maximum speed is approximately 360 m/min, or 6 m/s.

Exercise E

> 1 (a) For 3 minutes at 15 m/s the distance covered is 15×180 m, or 2.7 km, (b) during the $\frac{1}{2}$ min stop the distance covered is 0 km, and (c) for the next 4 min at 10 m/s the distance covered is 10×240 m, or 2.4 km (see Figure R).

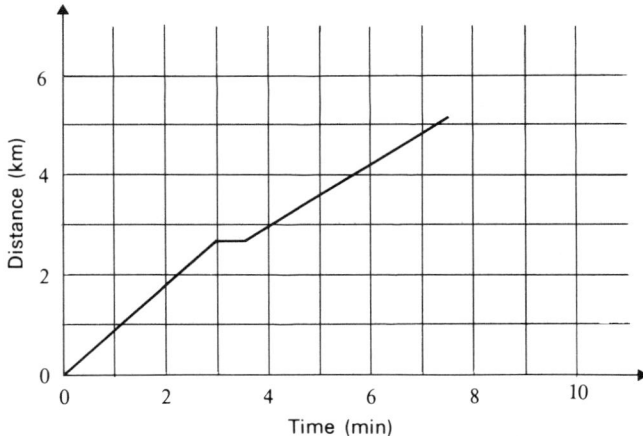

Figure R

2 (a)

t (s)	0	1	2	3	4
$s = 10t + t^2$ (m)	0	11	24	39	56

See Figure S over the page.

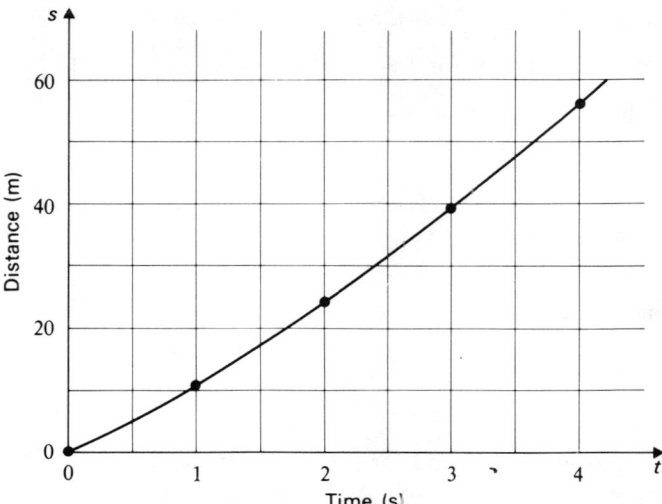

Figure S

(b) For $0 \leqslant t \leqslant 4$ the average speed is 14 m/s.
(c) 12 m/s, 14.8 m/s, 16 m/s.

Speed–time graphs

 1 The speed (a) increases steadily from 0 to 30 kph, (b) remains constant at 30 kph, (c) decreases steadily from 30 kph to 0 kph.
For $0 \leqslant t \leqslant 150$ the gradient is 0.2. It represents a rate of change of speed with respect to time of 0.2 kph/s.

 2 (a) The gradient is 0, and so there is no acceleration.
(b) The gradient is $^-0.6$, and so the acceleration is $^-0.6$ kph/s (or the deceleration is 0.6 kph/s).

 3 (a) It represents the acceleration, which decreases throughout the journey.
(b) The average gradient is 0.5 kph/s.
(c) The greatest acceleration (i.e. the steepest part of the graph) is where $t = 0$. It is approximately 2.5 kph/s.

Exercise F

 1 (a) Since the acceleration is constant, the gradient of the speed–time graph will be constant, i.e. the speed–time graph will be a straight line. As the acceleration is positive, the gradient of the speed–time graph will be positive. Hence (b) corresponds to the given acceleration–time graph.
(b) See Figure T.

 2 (a) The acceleration is $(20-0)/200$ m/s² or 0.1 m/s².
(b) The acceleration is $(24-15)/2$ m/s per minute, or (better) $(24-15)/120$ m/s², i.e. 0.075 m/s².
(c) The train is slowing down, so the *acceleration* $= {}^-0.25$ m/s², or the deceleration is 0.25 m/s².

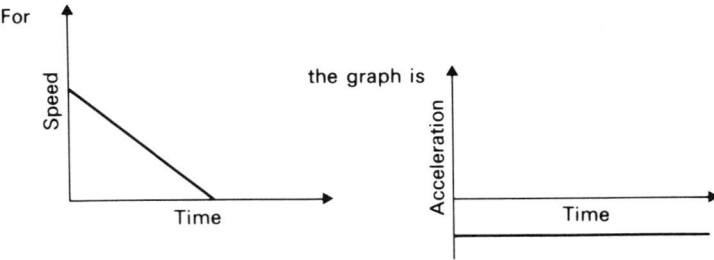

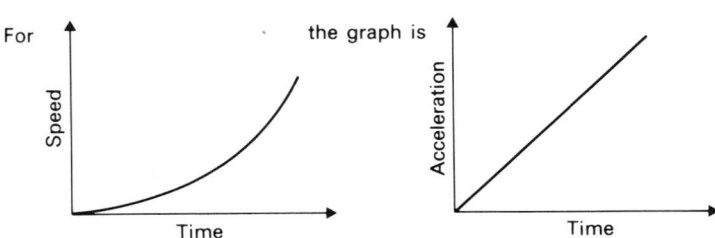

Figure T

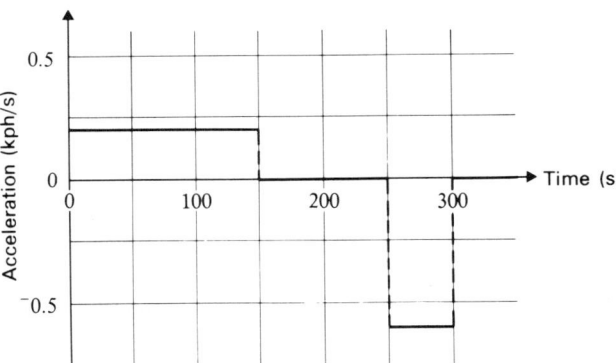

Figure U

3 See Figure U.

Post-test

> **1** The gradient of AB is $\frac{1}{4}$, of AC is $^-2$, of BC is 0, of CD is infinite, and of DB is $^-\frac{1}{3}$.
2 (a) $f: x \rightarrow 2+5x$. The gradient is 5.
(b) $g: x \rightarrow 3-8x$. The gradient is $^-8$.
(c) $4y+3x = 24$. The gradient is $^-\frac{3}{4}$.
These are plotted in Figure V over the page.
3 (a) Yes. For each increase in height of 1 km, the boiling point decreases by 3.1 °C, and so the gradient of the graph is constant, i.e. the graph will be a straight line (see Figure W over the page).
(b) The boiling point decreases at the rate of 3.1 °C/km. (i.e. the rate of change is $^-3.1$ °C/km.)

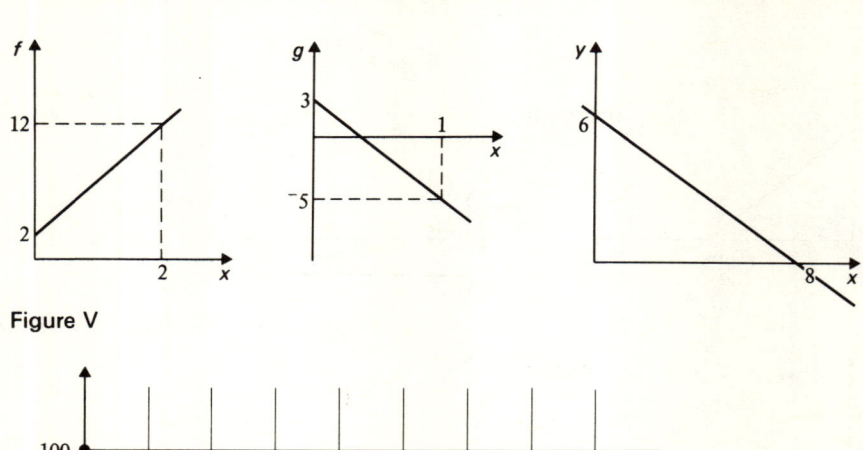

Figure V

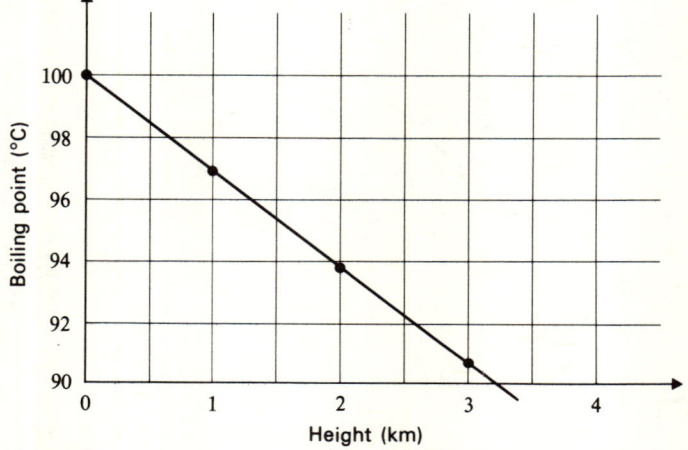

Figure W

(c) 100 °C − 85 °C = 15 °C fall in the boiling point.
15 °C/3.1 °C per km = 4.84 km (approximately). (15 900 ft).

4 By drawing tangents to the graph at the appropriate points the speeds are found to be approximately

Time (s)	1	2	3	4
Speed (m/s)	2	4	6	8

4 Linear programming

Objectives

This is what you should be able to do after studying this chapter.
(1) Represent the conditions of a problem by mathematical relations (often linear inequalities).
(2) Represent these relations graphically.
(3) Find the greatest (or least) value that an expression of the form $ax+by$ may take, subject to the particular restrictions of a problem (i.e. maximise – or minimise – a linear relation within a particular region.)

Pre-test

▷ 1 (a) From Figure 1, write down the relation that is satisfied by all points in the unshaded part of the plane, first, when the line $x+y = 6$ is drawn as a continuous line, and, secondly, when the line $x+y = 6$ is drawn as a dotted line.

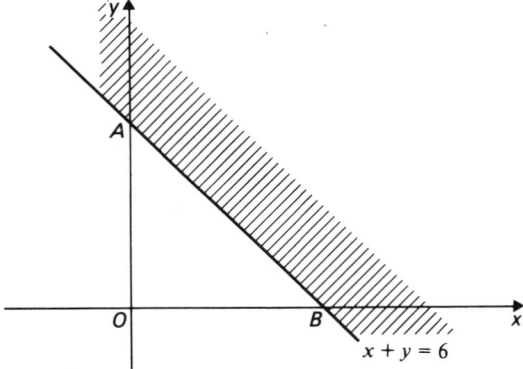

Figure 1

(b) Write down the coordinates of A and B.

2 What are the gradients of the lines (a) $2x+y = 7$ (b) $2x+y = {}^-4$ (c) $2x+y = 1$?
What is the form of the equation of a line parallel to $2x+y = 1$? How do the right-hand sides of these equations help you to 'put the lines in order'?

3   Show the following regions on separate sketch-graphs.
    (a) $2x - y > 6$ (b) $y \leqslant 3x$ (c) $x \geqslant 4$ (d) $y < 2x + 1$
    (Shade out the unwanted regions.)

4.1   Identifying the problem

What is the problem?

Firms producing cars always have a number of problems to face, quite apart from those that are of a political or personnel nature. In very simple terms, the Production Manager's problem is to fix a production cost and a selling price for each vehicle so that the firm is productive and profitable.

In order to achieve this aim he will have to consider such factors as the demand for a particular model of car, the number of workmen required, the time for each part of the process etc. Each of these conditions imposes a restriction upon the number of cars that can be made in a particular period of time.

In the final event, the firm will be concerned mainly with keeping the cost of production as low as possible, and making as large a profit as possible. We refer to these two conditions more concisely as: (1) minimising the cost of production, and (2) maximising the profit.

Nowadays a process called *linear programming* is used to help achieve these aims within the restrictions that apply.

The first stage is to represent the problem(s) and restrictions in mathematical terms, and in fact these problems and restrictions can often be represented by inequalities. This is known as 'making a mathematical model of the situation'.

The second stage is to put all these relations (or inequalities) together, and find possible solutions. In real life there may be up to 100 variables that have to be taken into account, and up to 100 inequalities or relations may well be needed: the only feasible method of solving these is by use of computers. In this chapter we shall consider problems that involve only two variables, so that the possible solutions may be obtained graphically.

The process of elimination

In many of our everyday activities we consciously or unconsciously use a process of elimination in selecting things that we want. Suppose you decide to buy a piece of recorded music with a gift token. Some of the decisions you have to make are the following.
(1)   Tape, cassette or disc?
(2)   How much extra can I afford?
(3)   What type of music do I want?
(4)   Do I want any particular performers?
By considering each condition in turn, you finally arrive at a 'short list' by eliminating a lot of the material in the shop as not being suitable to your particular requirements. This short list is the 'solution set' – there may still be several things to choose from; there may be only one that comes up to all the requirements, or there may be none!

> 1 Let us apply this to a simple mathematical example.
Start with the set of positive integers less than 20.
(a) List this set. (i.e. $I = \{x: 0 < x < 20\}$)
(b) By successive elimination, find the subset each of whose elements:
   (1) is not a multiple of 4
   (2) is a factor of 30
   (3) is not prime
   (4) is even
(c) Would the result be the same if you applied these conditions in another order?

Making the mathematical model

Now consider a slightly harder example. A game of dice is played by throwing one red die and one blue die together. To 'win' a throw, the following conditions have to be satisfied:
(1) The total score must be less than 8.
(2) The red score must not be greater than 4.
(3) The blue score must be greater than 2.
(4) The blue score must not be more than twice the red score.
What is the set of 'winning' throws? We could obviously find some by trial and error.

> 1 Are the following combinations winning throws?
(1) red score = 2, blue score = 4
(2) red score = 4, blue score = 2
(3) red score = 3, blue score = 5

and so on. Eventually we should find all the possible winning throws, but it is tedious, and we might miss out a few of the possible combinations that we ought to try.

We also find fairly soon that we could be more systematic. If we put condition (2) first, for example, we can eliminate quite a few combinations immediately. If we

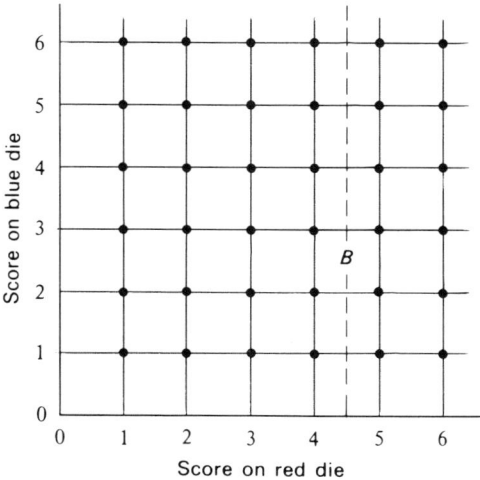

Figure 2

75

draw a diagram to show all the possible combinations (Figure 2) all those to the right of the line B are eliminated by condition (2).

It is now time to make our mathematical model. The diagram suggests that it would be reasonable to let the number on the red die be x, and the number on the blue die be y. Condition (2) therefore states $x \not> 4$.

2   Express this condition using (a) $<$ (b) $\leq$.
(c) Express the other three conditions similarly, using either $>$ or $<$.

Exercise A

1   A car firm has contracted to deliver at least 60 cars per day to Southampton, ready for export. The firm uses two types of loader; type A, which can carry 10 cars, and type B, which can carry 8 cars. The firm has 4 type A loaders, and 6 type B loaders, but only 8 drivers are available for the work. Each loader can make only one journey a day; it costs £100 a day to use a type A loader, and £60 a day to use a type B loader. Let the number of type A loaders used be x, and the number of type B loaders used be y.
(a) Write down two inequalities derived from the total number of each type of loader available.
(b) Write down the inequality that represents the condition on the total number of drivers available.
(c) If the firm uses 3 A loaders and 2 B loaders, what is the maximum number of cars that can be transported altogether?
(d) If it uses 1 A loader and 6 B loaders, how many cars can be carried?
(e) If it uses x A loaders and y B loaders, how many cars can be carried? Hence write down an inequality that states that at least 60 cars must be transported each day.
(f) How much does it cost to use 3 As and 2 Bs?
(g) How much does it cost to use 1 A and 6 Bs?
(h) How much does it cost to use x As and y Bs?
(i) If £P is the total cost when x As and y Bs are used, write down an expression for P in terms of x and y.

4.2 Graphical representation

1   Complete the following tables, and hence sketch the corresponding graphs.
(a) $x+2y = 12$,   (b) $3x+6y = 36$,   (c) $20x+40y = 240$.

x	0	
y		0

x	0	
y		0

x	0	
y		0

What do you notice?

Now let us return to the game of dice defined in the section called 'Making the mathematical model'.

2   Draw sketches to show the four orderings that we obtained for this problem namely
(a) $x+y < 8$ (b) $x < 5$ (c) $y > 2$ (d) $y < 2x+1$.

(Shade out the unwanted regions, and remember that, since we have used only $>$ and $<$, all the boundary lines are drawn as continuous lines to show that points on them are in the unwanted region; i.e. are not possible solutions.)

If we put these four graphs together we have the situation shown in Figure 3.

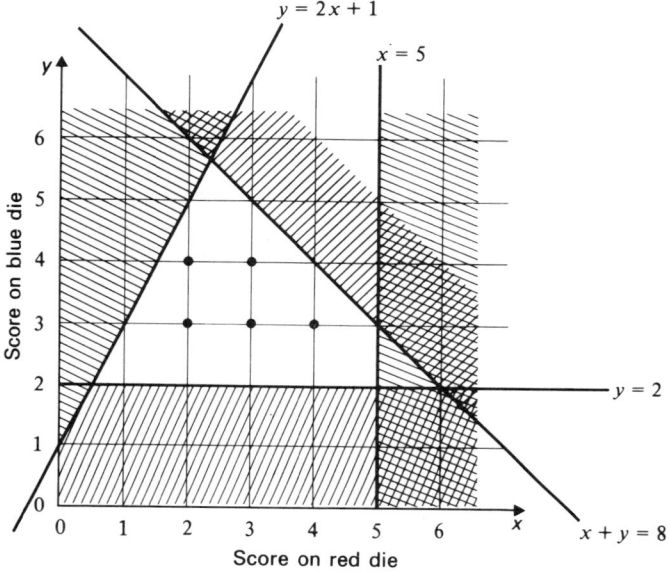

Figure 3

A 'winning' throw has to satisfy *all* the conditions laid down; i.e. must be represented in the *intersection* of the four sets that you have drawn above. In other words, the region in Figure 3 that is still unshaded contains *all* the possible winning combinations, and so we have now effectively solved the problem. The points ● show the five possible winning combinations.

▶ **3** Now show, on one graph, the four inequalities that you worked out for question **1** of Exercise A (parts (a), (b) and (e)).

Note that, as shown at the beginning of this section, equations that are basically the same have the same graph. Thus $10x + 8y \geq 60$ may be simplified to $5x + 4y \geq 30$ before we think about plotting it.

Write down the possible combinations that are solutions to the problem.

Exercise B

▶ **1** A hotel caters for school parties of not more than 30 people subject to the following conditions.
    (1) The minimum size for a party is 20 people.
    (2) The party must include at least 3 adults, but not more than 6 adults.
  (a) Using x to represent the number of children in a party, and y the number of adults, write down four inequalities connecting x and y.
  (b) Show these inequalities on a graph.

(c) How many adults can there be in a party that has 15 children?
(d) What is the maximum number of children that can be in a party with 4 adults?

▷ 2 A post office has to transport 900 parcels at one go using lorries that can take 150 parcels at a time, and vans that can take 60. The costs of each journey are £24 by lorry and £6 by van, and the total cost must be less than £120.
(a) Express these conditions as inequalities in x and y, where x is the number of lorries used and y is the number of vans used.
(b) Draw a graph to show the possible combinations.

4.3 Maximising and minimising

Once we have drawn the solution set for the 'two-dice game' problem we have effectively solved that problem; any one of the remaining combinations is a winning throw. But what about the last question in Exercise B?

So far, we have a number of possible combinations of lorries and vans; the final question might be

or
(a) Which is the most economical combination?
(b) Which combination employs the most drivers?

etc.

Suppose we wish to answer question (a). If the cost of using x lorries and y vans is £C; then $C = 24x + 6y$ and we want to make this cost as small as possible, i.e. we want to minimise C.

▷ 1 Sketch the graphs of $24x + 6y = C$ for $C = 30$, 60 and 90.

As in question 2 of the pre-test, we see that these lines are parallel, and, the larger the value of C, the further they move to the right. If we superimpose one of these (say for $C = 60$) on the graph we have so far drawn for this problem, we shall obtain Figure 4.

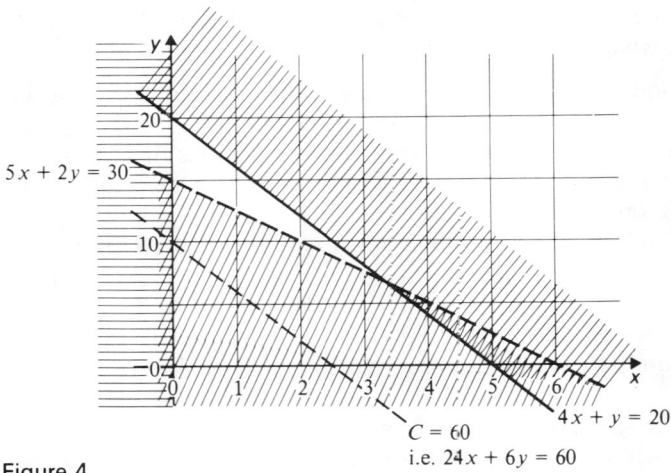

Figure 4

2 Now, mentally, move this line away from the origin, (but still keeping it parallel to its original direction) until it meets the unshaded region. Where does it first meet

the unshaded region? This point will represent the combination that makes C a minimum, and so is the answer to (a).

If, however, it was question (b) that had to be answered, then we need to consider the expression that tells us the total number of drivers used. This will be $(x+y)$ as each vehicle needs one driver.

> 3 On your graph, draw in one '$x+y$' line, e.g. $x+y = 10$. This time we want to maximise the quantity; and so as we move this line away from the origin (still keeping it parallel with its original direction) we want the *last* point of contact of the line with the unshaded region. Check that this is the point (0, 19). Why isn't it the point (0, 20)? Or the point (1, 19)? Or the point (0.2, 19)?

It is not always necessary to go through this fairly elaborate procedure, particularly if points *on* the boundary lines are permissible solutions. Suppose, in our example above, we had been allowed to spend up to *and including* £120 (so that points *on* the line $24x+6y = 120$ give permissible solutions), then it should be clear that the answers to both questions will be given by the coordinates of one of the corners of the unshaded region.

> 4 (a) What is the value of C (i.e. of $24x+6y$) for the points (0, 15), (0, 20) and $(3\frac{1}{3}, 6\frac{2}{3})$? Which of these gives the smallest value for C?
> (b) What is the value of $(x+y)$ at these three points? Which of these gives the largest number of drivers?

Exercise C

> 1 A toy firm makes two kinds of toy soldiers, using a machine that can work for 10 hours a day. The 'Guard' takes 8 s to make, and contains 8 g of metal. The 'Cavalryman' is made in 6 s, and uses 16 g of metal. 64 kg of metal is available each day.

If the profit on the 'Guard' is 5p, and that on the 'Cavalryman' is 6p, how many of each should be made per day to maximise the profit?

> 2 Ten men are available to unload seven lorries, but not more than two men can work on any one lorry. If x is the number of lorries being unloaded by one man, and y the number being unloaded by two men, write down two relations (other than $x \geqslant 0$ and $y \geqslant 0$) satisfied by x and y, and show them on a graph.

Each lorry carries one tonne of goods, and experience shows that two men working together can unload three times as fast as one man by himself. If one man on his own can unload 0.2 tonnes per hour, show that the initial rate of working is $\frac{1}{10}(2x+6y)$ tonnes per hour. Use your graph to find the values of x and y that will give the quickest initial rate of working.

Summary

(1) Many problems can be solved by forming a 'mathematical model' to represent the conditions that the problem poses. This 'model' is often a number of inequalities (or orderings) in two or more unknowns (or variables).
(2) If only two variables are involved in the model, then a solution can be found graphically by plotting the various inequalities.

(3) For given values of a and b the various lines obtained from
$$ax + by = k$$
by giving different values to k form a set of parallel lines.

The line $ax + by = 0$ goes through the origin, and as k increases the lines move further to the right (see Figure 5).

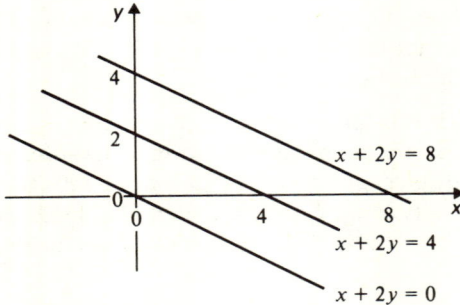

Figure 5

(4) The ideas of the previous paragraph are used to maximise (or minimise), within the solution region of the graph, quantities that can be expressed in the form $ax + by$.

Post-test

1. What graphs would you draw to represent the following regions? Would you draw them as dotted lines or continuous lines (assuming that you will be shading out the unwanted regions)? Would the shading be *above* the line, or below it (or, if this is impossible to state, to the left or right)?
   (a) $x \geqslant 0$ (b) $y > 0$ (c) $2x + y > 4$ (d) $x + 3y \geqslant 9$ (e) $x + y < 6$

2. (a) Draw x- and y-axes from 0 to 9 and show, on one graph, the five regions listed in question 1.
   (b) If all the inequalities are to be taken simultaneously, how many solutions are there if x and y are integers?
   (c) How many are there if x and y are real numbers?
   (d) What is the greatest and the least value that x can take, subject to these conditions, if x is an integer?
   (e) What is the greatest and least value of y, if y is an integer?

3. For a camp of 70 children, two types of tent are available on hire. The Patrol tent sleeps 7 and costs £10 a week to hire; the Hike tent sleeps two and costs £2 a week. The total number of tents must not exceed 19.

   Write down two inequalities connecting the number of Patrol tents (p) and the number of Hike tents (h), and an expression for the cost of hiring these numbers of tents for a week.

   By a graphical method, or otherwise, find the most economical cost of hire, and the number of each type of tent required in this case. (For a graphical method, use a scale of 1 cm to 2 tents.)

Assignment

1 (a) On one graph show clearly the region R defined by the following inequalities. (Shade out the unwanted regions)

$$y \leq 8, \quad y \leq 2x, \quad 5x+2y \leq 51, \quad x+2y \geq 15$$

   (b) Within the region R, state the maximum value of $x+y$, the minimum value of $10x+y$, and the maximum value of $x-y$.

   (c) Repeat part (b) when the boundary lines are *excluded* from R, and x and y can take integral values only.

2 A contractor who hires earth-moving equipment has the choice of two types of machine. Type A costs £25 per day to hire, needs one man to operate it, and moves 30 tonnes of earth per day. Type B costs £10 per day, needs four men to operate it, and moves 70 tonnes of earth per day.

   The contractor has a labour force of 64 men available, can use a maximum of 25 machines on the site, and is allowed to spend up to £500 per day on the operation.

   (a) If he hires x machines of type A and y machines of type B, write down three inequalities (other than $x \geq 0$ and $y \geq 0$) which must be satisfied. Show graphically the set of possible values of (x, y).

   (b) Write down an expression, in terms of x and y, for the number of tonnes of earth moved daily, and hence find the maximum weight of earth that the contractor can move in one day.

Answers

Pre-test

> 1 (a) A continuous line implies that points on the line belong to the shading. Hence the unshaded region is $x+y < 6$. When a dotted line is drawn, points on the line now belong to the unshaded region, which is therefore $x+y \leq 6$.

   (b) A is the point $(0, 6)$ and B is the point $(6, 0)$.

2 (a) Rewriting the equation as $y = 7-2x$ shows that the gradient is -2.
   (b) and (c) also have a gradient of -2.
   All lines parallel to $2x+y = 1$ will have an equation of the form $2x+y = k$, for various values of k. The larger the value of k, the further to the right the lines are, as shown in Figure A over the page.

3 The answers are shown in Figure B over the page.

4.1 Identifying the problem

The process of elimination

> 1 (a) $I = \{1, 2, 3, 4, 5, 6, 7, 8, 9, 10, 11, 12, 13, 14, 15, 16, 17, 18, 19\}$.

81

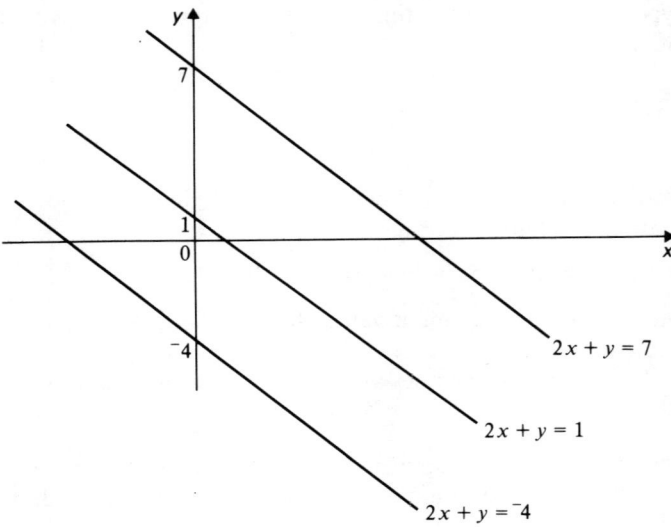

Figure A

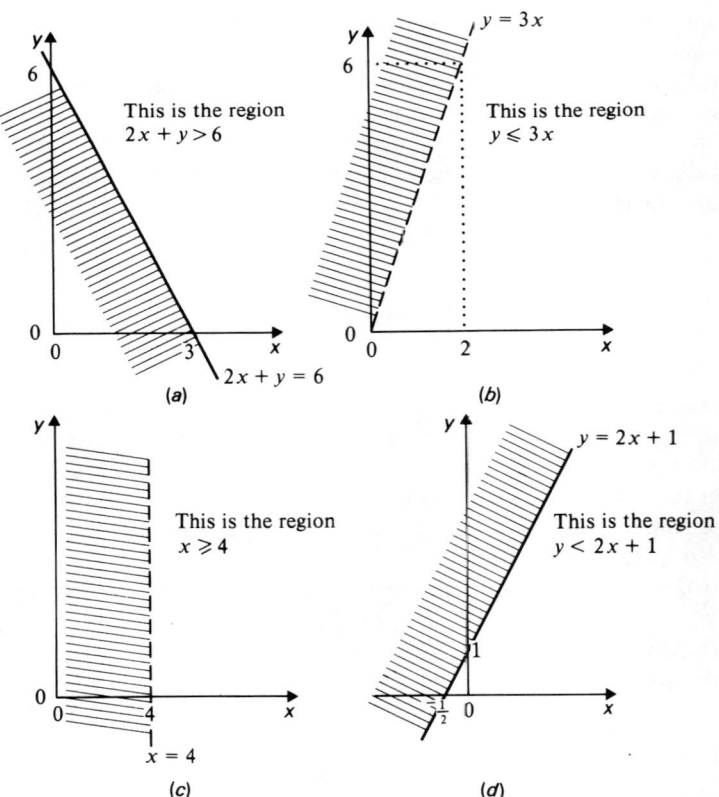

Figure B

(b) First 4, 8, 12, 16 are eliminated, then (4), 7, (8), 9, 11, (12), 13, 14, (16), 17, 18 and 19 are eliminated. (So at this stage we have only 1, 2, 3, 5, 6, 10 and 15 left.) Now 2, 3 and 5 are eliminated (as they are prime), and 1 and 15 are eliminated (as they are not even), and we are left with the subset {6, 10}.

(c) The order in which these conditions are applied does not affect the final result.

Making the mathematical model

▷1 red = 2 and blue = 4 is a winning combination,
red = 4, blue = 2 is not, as condition (3) is not satisfied,
red = 3, blue = 5 is not, as condition (1) is not satisfied.

▷2 $x \not> 4$ can be written as (a) $x < 5$ or (b) $x \leq 4$.
The first of these is true only if x can take integral values only, whereas the second form is correct for all real values of x.

(1) can be written $x+y < 8$, and (3) is $y > 2$. 'Literally' (4) is $y \not> 2x$, i.e. $y \leq 2x$, and since x and y can take integral values only, this can be written as $y < 2x+1$.

Exercise A

▷1 (a) As there are 4 type A loaders, $x \leq 4$, and 6 type B loaders, $y \leq 6$.
(b) We assume that each loader requires only one driver, so that x type A loaders require x drivers, and y type B loaders require y drivers, and, if x and y are the number of loaders actually in use, $x+y \leq 8$.
(c) 3 A loaders can carry a maximum of $3 \times 10 = 30$ cars, and 2 B loaders can carry a maximum of 2×8 cars; a total of 46 cars (maximum).
(d) A maximum total of $(1 \times 10)+(6 \times 8) = 58$ cars.
(e) A maximum of $10x+8y$ cars.
The firm is required to deliver *at least* 60 *cars* each day, and so the capacity of the loaders actually used must be at least 60. (The number of 'car spaces' available will be *more than* 60 if not all the loaders are full.) The inequality is therefore

$$10x+8y \geq 60.$$

(f) The cost of 3 As and 2 Bs is £$(3 \times 100)+(2 \times 60) = £420$.
(g) The cost of 1 A and 6 Bs is £460.
(h) The cost of x As and y Bs is £$(100x+60y)$.
(i) $P = 100x+60y$, from (h).

4.2 Graphical representation

▷1 (a) $x+2y = 12$ (b) $3x+6y = 36$ (c) $20x+40y = 240$

x	0	12
y	6	0

x	0	12
y	6	0

x	0	12
y	6	0

They all have the same graph, as shown in Figure C over the page; i.e. they are 'equivalent equations'. (Compare the idea of equivalent fractions $\frac{1}{2}$, $\frac{2}{4}$, $\frac{5}{10}$ etc.)

▷2 The answers are shown in Figure D over the page.

▷3 The result is shown in Figure E over the page.

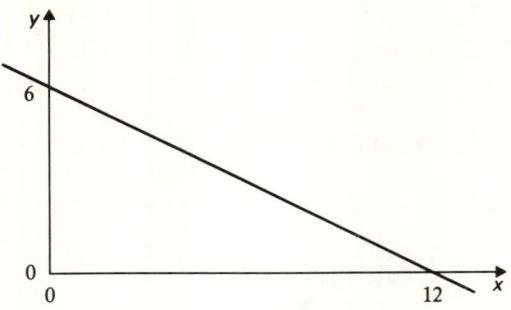

Figure C

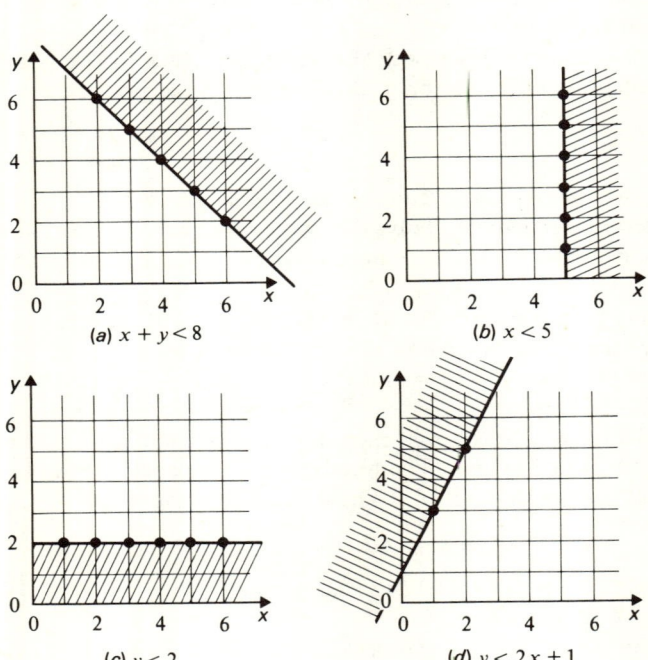

Figure D

The boundary for $x+y \leq 8$ is $x+y = 8$, which goes through the points (0, 8) and (8, 0). The boundary for $5x+4y \geq 30$ is $5x+4y = 30$. When y is 0, $5x = 30$ or $x = 6$, so one point on the line is (6, 0). Similarly $(0, 7\frac{1}{2})$ is the point where the line cuts the y-axis. As all the inequalities are of the form $A \geq B$, all the points on the boundary lines are possible solutions. Hence possible solutions are (2, 5), (2, 6), (3, 4), (3, 5), (4, 3), (4, 4).

Exercise B

 1 (a) The hotel cannot take more than 30 people: $x+y \leq 30$
the minimum size of a party is 20 people: $x+y \geq 20$

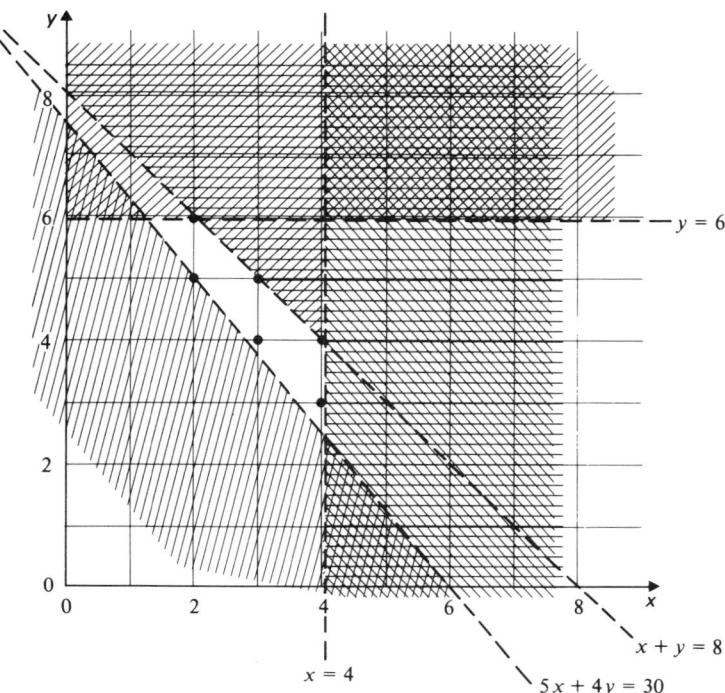

Figure E

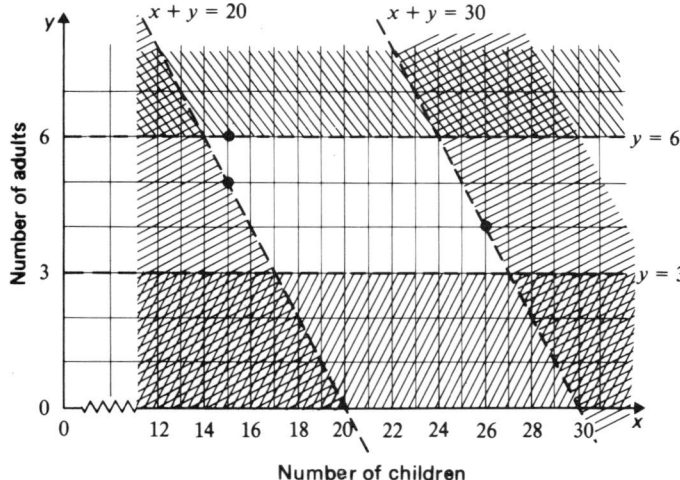

Figure F

there must be a minimum of 3 adults: $y \geqslant 3$
and a maximum of 6 adults: $y \leqslant 6$
(b) These inequalities are shown on Figure F.
(c) If $x = 15$, then y can be 5 or 6.
(d) If $y = 4$, then the maximum value of x is 26.

2 (a) There are 900 parcels to be carried: $150x+60y \geq 900$;
total cost must be less than £120: $24x+6y < 120$.
(Note that the first condition means that there must be enough space available to carry *at least* 900 parcels; i.e. there must be a *minimum* of 900 'parcel-spaces' on the x lorries and y vans so that the minimum value of $150x+60y$ is 900.)
These two inequalities simplify to $5x+2y \geq 30$, and $4x+y < 20$.
(b) These two inequalities are plotted in Figure G.

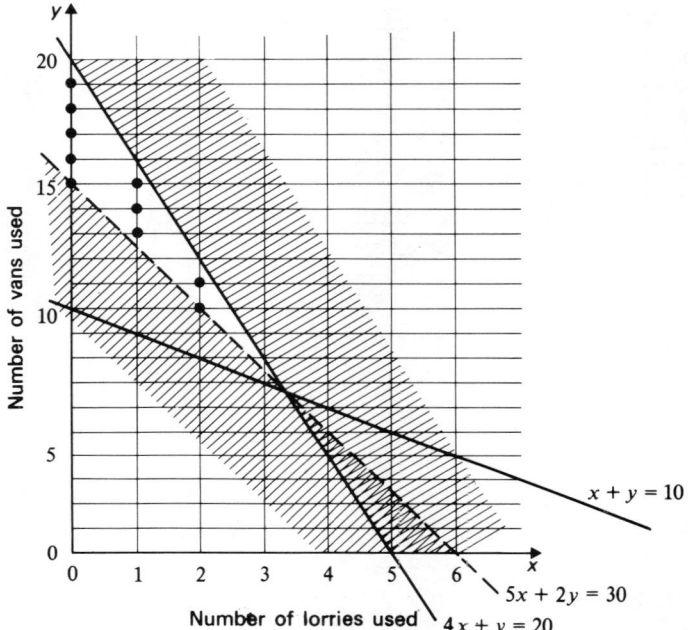

Figure G

Possible combinations are shown by the dots.

4.3 Maximising and minimising

1 The graphs are shown in Figure H.

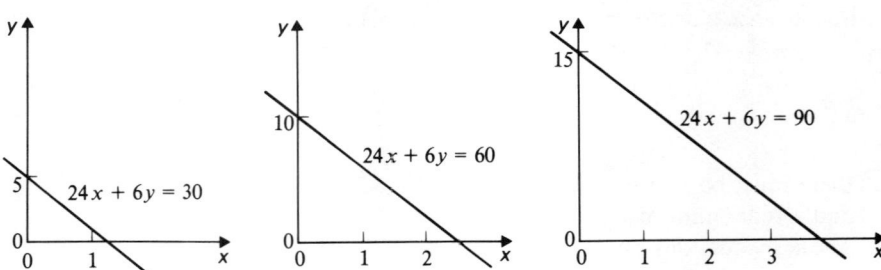

Figure H

2 The line $24x + 6y = C$ first meets the unshaded region in Figure 4 at the point (0, 15). At this point the value of C is $0 + 6 \times 15 = 90$; and so the minimum cost is £90.

3 The total cost must be *less than* £120, and so the combinations (0, 20) and (1, 19) are not possible as they give costs of exactly £120, and £138 respectively. Points with non-integral coordinates are not permissible as x and y refer to the number of vans and lorries, and these must be whole numbers! Hence the maximum number of drivers that can be used, subject to the total cost condition, is 19.

4

Point	(0, 15)	(0, 20)	$(3\frac{1}{3}, 6\frac{2}{3})$	(3, 8)	(2, 10)
Value of C (i.e. $24x + 6y$)	90	120	120	120	108
Value of $(x + y)$	15	20	10	11	12

As the point $(3\frac{1}{3}, 6\frac{2}{3})$ is not a lattice point (i.e. a point whose coordinates are integers) we have included the two lattice points in the solution set nearest to it.
(a) The minimum value of C is 90 (i.e. minimum cost is £90) which occurs when $x = 0$ and $y = 15$.
(b) The maximum value of $(x + y)$ is 20, occurring when $x = 0$ and $y = 20$.

Exercise C

1 Maximum machine time is 10 hours: $8x + 6y \leqslant 36000$.
(The time taken to produce x 'Guards' and y 'Cavalrymen' must be less than 10 hours.)
Metal available is 64 kg: $8x + 16y \leqslant 64000$.
The profit £P is given by $P = (5x + 6y)/100$, and this is to be maximised.
Let $x = 1000X$ and $y = 1000Y$; then the two inequalities simplify to $4X + 3Y \leqslant 18$, and $X + 2Y \leqslant 8$; and $P = 10(5X + 6Y)$.
The result of plotting these inequalities is shown in Figure I over the page. [Point of intersection of lines is Q (2.4, 2.8).] The line $P = 100$ (i.e. $5X + 6Y = 10$) has been drawn in the bottom left-hand corner; this represents a profit of £100. To find the maximum profit we must move this line away from the origin (keeping it parallel to its present direction, i.e. in the direction of the arrows) until we reach the 'last' point in the solution set. This point will represent the combination that gives the maximum profit.
By inspection, this is the point Q, but, if this is not clear, evaluate the profit for the three points (0, 4), Q and $(4\frac{1}{2}, 0)$. These profits will be £240, £288 and £225 respectively (using $P = 10(5X + 6Y)$, and so the firm should make 2400 'Guards' and 2800 'Cavalrymen' per day, which will give a profit of £288.
(The substitutions $x = 1000X$ and $y = 1000Y$ are not essential, but they do simplify the arithmetic a little. What we are saying is that the factory produces X thousand 'Guards' and Y thousand 'Cavalrymen' each day.)

2 Ten men are available: $x + 2y \leqslant 10$.
Seven lorries are to be unloaded: $x + y \leqslant 7$.
The result of plotting these inequalities is shown in Figure J over the page. One man on his own unloads at the rate of 0.2 tonnes per hour. Hence two men working together unload at the rate of 0.6 tonnes per hour. The rate of unloading

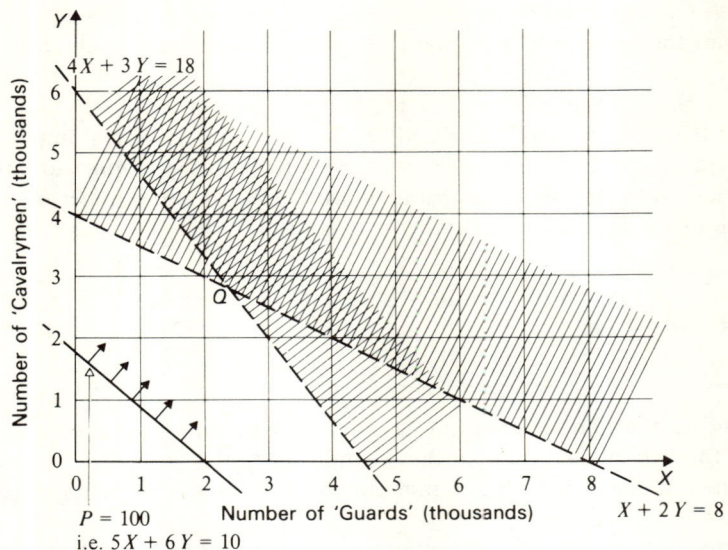

Figure I

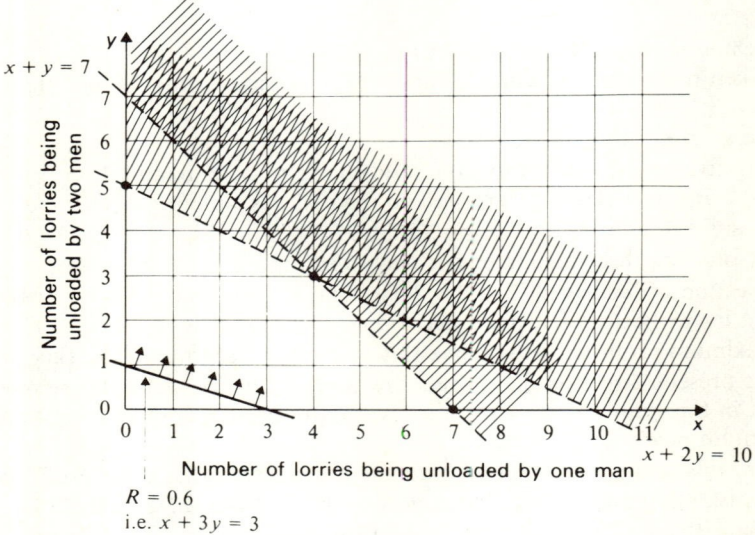

Figure J

is thus $0.2x + 0.6y$ tonnes per hour. If $R = 0.2x + 0.6y$, then the line $R = 0.6$ is that shown on the graph. Within the solution set, R is a maximum at $(0, 5)$, giving a maximum rate of working of 3 tonnes per hour.

Post-test

1

	Region	Boundary	Type of line	Shading
(a)	$x \geqslant 0$	$x = 0$ (y-axis)	--------	to the left
(b)	$y > 0$	$y = 0$ (x-axis)	————	below the line
(c)	$2x+y > 4$	$2x+y = 4$	————	below (and to the left)
(d)	$x+3y \geqslant 9$	$x+3y = 9$	--------	below (and to the left)
(e)	$x+y < 6$	$x+y = 6$	————	above (and to the right)

2 (a) The result is shown in Figure K.

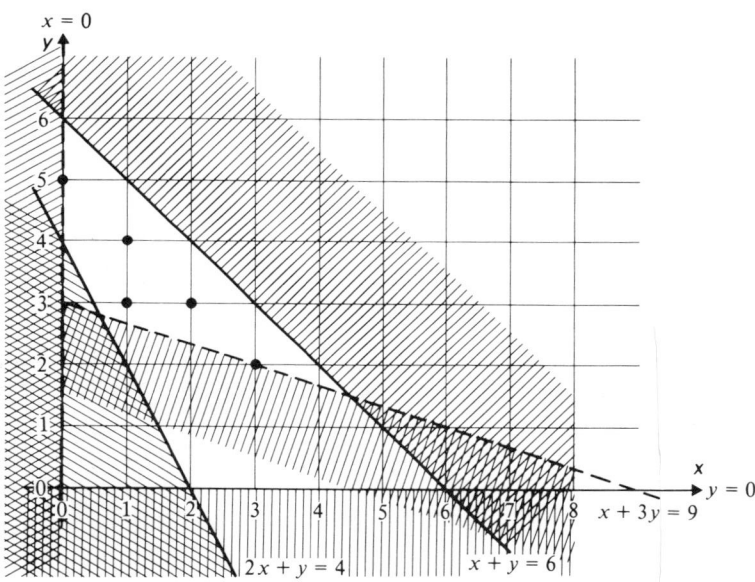

Figure K

(b) If x and y are integers, there are 5 possible solutions, shown by the heavy lattice points above. ((0, 5), (1, 3), (1, 4), (2, 3) and (3, 2).)

(c) If x and y can be any real numbers then there is an infinite number of solutions. (Any point in the unshaded region is allowed.)

(d) From the solutions listed above in part (b) the maximum (integral) value of x is 3, and the minimum value is 0.

(e) Also, the maximum integral value of y is 5, and the minimum is 2.

3 70 children have to be accommodated: $7p+2h \geqslant 70$; the maximum number of tents is 19: $p+h \leqslant 19$.

The cost of hiring these tents for one week (£C) is given by $C = 10p+2h$.

Plotting the two inequalities gives Figure L over the page. The 'cost line' $C = 20$ is shown at the bottom left-hand corner. As the cost-lines are 'moved across' the page, the first lattice point (in the solution set) that is reached is (7, 11); this, therefore, represents the most economical arrangement, namely 7 Patrol tents and 11 Hike tents at a total cost per week of £92. (As a check, the combination (7, 12) costs

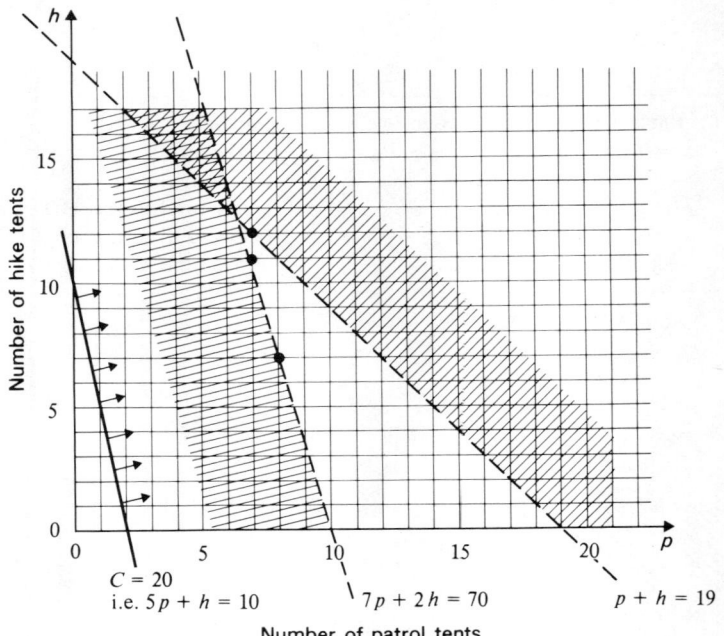

Figure L

£94, and the combination (8, 7) costs £94 as well. It is also worth checking that the points (7, 10) and (6, 13) are definitely to the *left* of the line $7p+2h = 70$, and so are not in the solution set.)

Published by the Press Syndicate of the University of Cambridge
The Pitt Building, Trumpington Street, Cambridge CB2 1RP
32 East 57th Street, New York, NY 10022, USA
296 Beaconsfield Parade, Middle Park, Melbourne 3206, Australia

© Cambridge University Press 1981

First published 1981

Printed in Great Britain at the
University Press, Cambridge

British Library cataloguing in publication data
School Mathematics Project
Individualised mathematics.
Algebra 2: Equations, formulas and graphs
1. Mathematics – 1961–
I. Title   II. National Extension College
510   QA39.2   80-49964
ISBN 0 521 24332 8